6600 Practice Problems

100+ Days of Timed Tests

Double Digit
Addition & Subtraction

Grade 1-3

60

120 Pages

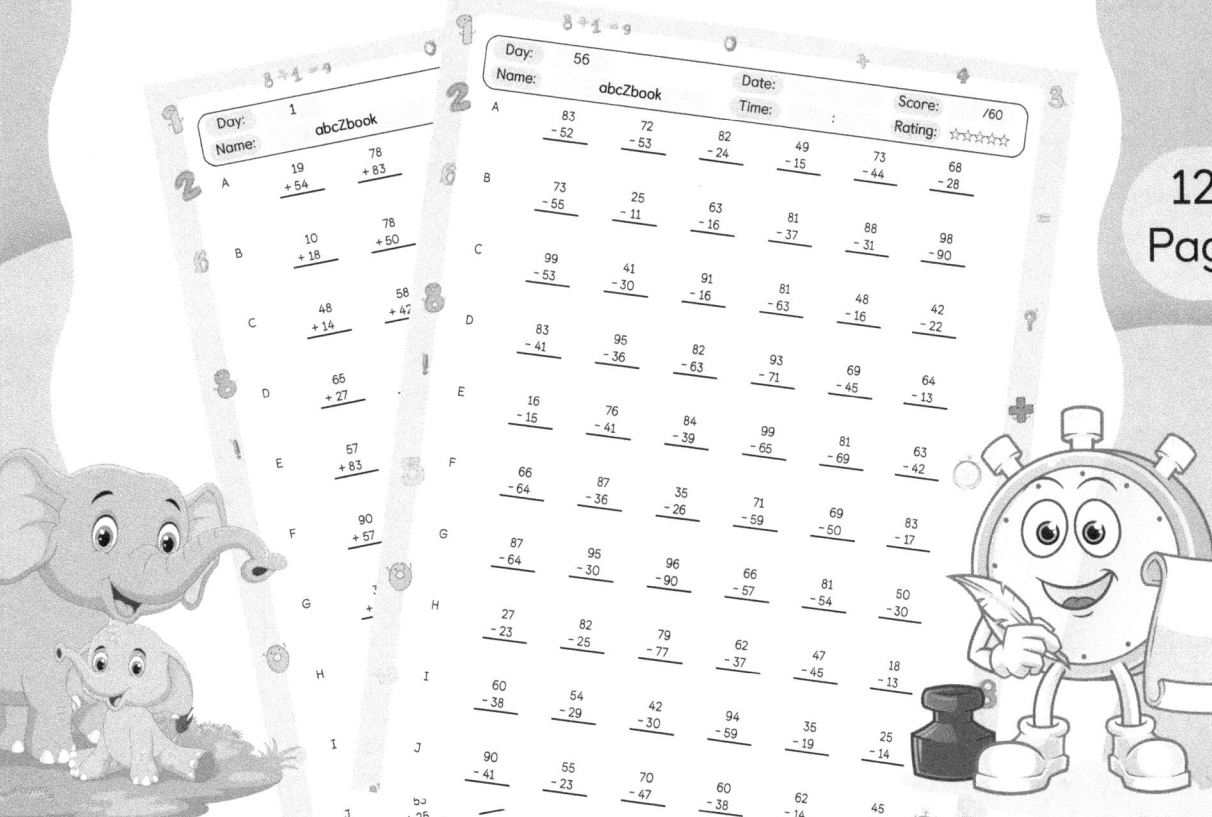

Ages 6-9

MATH DRILLS WITH & WITHOUT REGROUPING

Dear Parents,

Thank you for your purchase!

I sincerely hope, this book will be more helpful and interesting for the kids to learn **Double Digit Addition & Subtraction.**

Your opinion matters to us. We'd love to hear from you! You can leave your valuable comments and feedbacks on Amazon. Please take a moment to write a review. We really appreciate your support.

Email us at abczbook@gmail.com with title "**Double Digit Addition & Subtraction**" and get your free practice worksheets!

Hopefully, your kids will enjoy this book.

Enjoy Learning!

abcZbook Press

abcZbook Press

This book belongs to

A few minutes of math workout everyday will help the children master the math skills.

This **"Double Digit Addition & Subtraction: 100+ Days of Timed Tests"** is the beginner's level math practice workbook for Grade 1 to 3.

This book is specifically designed for numbers upto 99 (double digit). These set of math practice worksheets are designed to test addition and subtraction with and without regrouping (or carrying). The kids can challenge themselves with the timed test problems. This book mainly focuses on improving both addition & subtraction skills and building confidence level.

This book also has Answer Key sheets at the end of the book, so that you can easily check with the kid's answer.

In this book, there are 60 problems to be solved on a daily basis and total of 110 pages of Timed test practice sheets. It helps the kids to perform consistently and trained to be excellence in addition and subtraction.

It also has Bonus pages to get the kids ready for the higher number addition and subtraction.

Please check out other books without regrouping.

Table of Contents:

A
| 19 | 78 | 55 | 12 | 58 | 11 |
| + 54 | + 83 | + 17 | + 14 | + 17 | + 94 |

B
| 10 | 78 | 21 | 35 | 34 | 84 |
| + 18 | + 50 | + 50 | + 94 | + 20 | + 11 |

C
| 48 | 58 | 96 | 15 | 76 | 44 |
| + 14 | + 42 | + 13 | + 71 | + 84 | + 29 |

D
| 65 | 74 | 81 | 64 | 56 | 72 |
| + 27 | + 71 | + 54 | + 90 | + 69 | + 46 |

E
| 57 | 50 | 90 | 33 | 99 | 42 |
| + 83 | + 66 | + 24 | + 33 | + 91 | + 34 |

F
| 90 | 34 | 20 | 90 | 14 | 79 |
| + 57 | + 94 | + 51 | + 49 | + 63 | + 75 |

G
| 34 | 94 | 45 | 36 | 45 | 37 |
| + 73 | + 26 | + 75 | + 41 | + 49 | + 65 |

H
| 92 | 52 | 69 | 64 | 21 | 33 |
| + 40 | + 11 | + 84 | + 65 | + 79 | + 97 |

I
| 67 | 39 | 29 | 60 | 28 | 77 |
| + 77 | + 59 | + 61 | + 50 | + 34 | + 38 |

J
| 53 | 73 | 63 | 29 | 97 | 77 |
| + 25 | + 39 | + 79 | + 84 | + 98 | + 42 |

A

20	46	72	72	18	76
+ 83	+ 55	+ 88	+ 93	+ 38	+ 32

B

77	10	17	89	98	26
+ 83	+ 97	+ 46	+ 92	+ 58	+ 45

C

18	25	12	36	65	41
+ 76	+ 77	+ 72	+ 99	+ 38	+ 11

D

23	76	25	60	54	91
+ 59	+ 79	+ 55	+ 54	+ 62	+ 71

E

13	85	64	27	29	37
+ 49	+ 71	+ 71	+ 62	+ 39	+ 66

F

74	65	97	94	49	75
+ 31	+ 49	+ 92	+ 30	+ 62	+ 47

G

18	85	68	29	49	20
+ 61	+ 61	+ 26	+ 61	+ 83	+ 48

H

43	66	64	74	39	92
+ 60	+ 52	+ 56	+ 60	+ 64	+ 50

I

93	98	89	25	99	96
+ 14	+ 69	+ 54	+ 18	+ 28	+ 74

J

57	11	89	66	10	21
+ 19	+ 62	+ 22	+ 25	+ 60	+ 28

A	70 + 57	54 + 71	71 + 15	34 + 39	27 + 49	79 + 10
B	35 + 55	31 + 18	42 + 64	77 + 17	57 + 63	10 + 31
C	40 + 79	25 + 23	86 + 94	80 + 88	43 + 53	91 + 44
D	60 + 97	20 + 35	20 + 96	27 + 77	63 + 57	24 + 59
E	97 + 89	30 + 62	50 + 29	89 + 79	41 + 74	15 + 24
F	81 + 99	49 + 66	69 + 60	15 + 25	62 + 32	44 + 60
G	73 + 77	66 + 25	55 + 14	43 + 84	32 + 88	73 + 15
H	17 + 84	72 + 29	21 + 42	31 + 29	66 + 96	76 + 95
I	45 + 54	15 + 19	33 + 93	55 + 49	95 + 46	85 + 39
J	80 + 90	77 + 34	84 + 90	67 + 61	85 + 87	36 + 76

30

60

A.
75	73	34	20	77	89
+ 39	+ 62	+ 32	+ 64	+ 82	+ 71

B.
99	23	20	34	26	30
+ 80	+ 55	+ 29	+ 71	+ 14	+ 49

C.
15	39	62	16	74	68
+ 30	+ 99	+ 46	+ 28	+ 49	+ 68

D.
61	18	42	38	65	86
+ 85	+ 26	+ 79	+ 56	+ 37	+ 52

E.
80	76	38	75	74	33
+ 40	+ 30	+ 73	+ 37	+ 22	+ 57

F.
29	56	56	60	22	38
+ 26	+ 44	+ 24	+ 18	+ 80	+ 74

G.
94	56	41	95	45	26
+ 86	+ 62	+ 94	+ 93	+ 23	+ 78

H.
67	24	61	16	64	62
+ 13	+ 41	+ 47	+ 92	+ 17	+ 60

I.
30	48	79	13	29	24
+ 57	+ 94	+ 59	+ 17	+ 69	+ 23

J.
43	61	41	90	23	14
+ 72	+ 95	+ 28	+ 82	+ 92	+ 37

A
| 51 | 69 | 18 | 91 | 23 | 49 |
| + 21 | + 59 | + 29 | + 42 | + 62 | + 16 |

B
| 86 | 48 | 56 | 28 | 25 | 40 |
| + 37 | + 16 | + 87 | + 43 | + 42 | + 35 |

C
| 52 | 30 | 68 | 95 | 36 | 80 |
| + 83 | + 38 | + 57 | + 99 | + 91 | + 60 |

D
| 88 | 96 | 56 | 90 | 24 | 70 |
| + 10 | + 72 | + 97 | + 87 | + 39 | + 29 |

E
| 18 | 46 | 31 | 75 | 36 | 25 |
| + 49 | + 98 | + 38 | + 48 | + 89 | + 63 |

F
| 26 | 93 | 60 | 62 | 90 | 17 |
| + 12 | + 80 | + 72 | + 48 | + 60 | + 54 |

G
| 82 | 75 | 29 | 42 | 69 | 51 |
| + 10 | + 80 | + 84 | + 42 | + 22 | + 75 |

H
| 28 | 87 | 82 | 35 | 68 | 58 |
| + 99 | + 36 | + 18 | + 53 | + 61 | + 70 |

I
| 44 | 83 | 45 | 23 | 98 | 15 |
| + 24 | + 71 | + 17 | + 87 | + 87 | + 62 |

J
| 38 | 36 | 11 | 40 | 25 | 18 |
| + 89 | + 69 | + 47 | + 57 | + 63 | + 80 |

A	13 + 98	39 + 75	28 + 80	65 + 95	71 + 66	64 + 92
B	97 + 53	59 + 61	33 + 28	63 + 76	56 + 95	28 + 13
C	24 + 71	50 + 53	67 + 52	98 + 88	62 + 69	82 + 85
D	61 + 85	89 + 97	15 + 78	37 + 18	71 + 65	23 + 51
E	66 + 37	79 + 38	53 + 23	87 + 57	68 + 84	29 + 72
F	86 + 39	56 + 66	89 + 98	73 + 22	94 + 82	19 + 42
G	13 + 48	53 + 14	72 + 31	78 + 16	34 + 49	56 + 28
H	53 + 19	97 + 13	67 + 66	49 + 67	70 + 80	78 + 65
I	53 + 87	31 + 98	64 + 18	55 + 48	56 + 82	76 + 19
J	18 + 49	90 + 19	80 + 51	63 + 36	12 + 61	79 + 11

30

60

A

35	17	24	49	96	66
+ 15	+ 63	+ 68	+ 92	+ 67	+ 58

B

72	28	18	83	53	68
+ 58	+ 26	+ 20	+ 89	+ 14	+ 11

C

79	37	17	75	13	10
+ 97	+ 97	+ 15	+ 26	+ 79	+ 54

D

21	21	89	69	78	46
+ 53	+ 70	+ 66	+ 64	+ 78	+ 37

E

87	81	74	10	59	52
+ 99	+ 20	+ 64	+ 27	+ 68	+ 71

F

81	65	53	95	81	13
+ 55	+ 37	+ 55	+ 64	+ 13	+ 70

G

98	32	49	67	12	14
+ 49	+ 77	+ 37	+ 96	+ 90	+ 88

H

39	23	15	59	65	26
+ 18	+ 66	+ 24	+ 33	+ 51	+ 66

I

45	45	88	33	74	74
+ 92	+ 71	+ 43	+ 12	+ 94	+ 47

J

22	23	60	87	97	73
+ 82	+ 10	+ 46	+ 75	+ 51	+ 23

A	16 + 88	40 + 35	61 + 26	71 + 58	77 + 21	59 + 70
B	38 + 98	75 + 63	60 + 55	39 + 41	99 + 35	36 + 98
C	32 + 29	38 + 71	61 + 32	14 + 93	35 + 38	34 + 29
D	52 + 79	52 + 77	77 + 96	23 + 42	97 + 76	33 + 63
E	97 + 85	59 + 55	29 + 23	39 + 22	10 + 11	27 + 41
F	18 + 26	73 + 90	25 + 76	61 + 41	85 + 26	29 + 10
G	41 + 79	68 + 47	93 + 19	17 + 27	15 + 72	25 + 73
H	10 + 92	12 + 42	44 + 98	65 + 47	95 + 80	84 + 40
I	45 + 67	16 + 12	70 + 68	98 + 90	76 + 38	39 + 72
J	83 + 33	14 + 24	25 + 85	44 + 19	30 + 76	63 + 22

A

55	94	75	88	99	66
+ 68	+ 37	+ 95	+ 38	+ 12	+ 67

B

68	91	88	66	68	86
+ 19	+ 95	+ 64	+ 86	+ 82	+ 35

C

84	16	76	72	41	66
+ 16	+ 92	+ 29	+ 15	+ 61	+ 56

D

11	26	17	35	98	41
+ 59	+ 41	+ 29	+ 85	+ 86	+ 89

E

35	16	63	81	98	13
+ 89	+ 72	+ 43	+ 84	+ 57	+ 94

F

14	43	93	33	19	19
+ 37	+ 69	+ 71	+ 50	+ 59	+ 87

G

37	55	93	81	12	60
+ 46	+ 10	+ 40	+ 26	+ 15	+ 57

H

24	83	41	38	75	84
+ 72	+ 69	+ 23	+ 78	+ 72	+ 71

I

41	82	56	94	61	86
+ 52	+ 53	+ 28	+ 85	+ 21	+ 14

J

89	25	32	79	23	54
+ 81	+ 51	+ 86	+ 60	+ 42	+ 89

A

10	75	90	16	40	24
+ 51	+ 20	+ 83	+ 30	+ 40	+ 65

B

88	66	86	48	15	42
+ 95	+ 41	+ 97	+ 69	+ 43	+ 48

C

76	35	85	58	76	65
+ 75	+ 18	+ 54	+ 20	+ 73	+ 85

D

35	46	81	98	10	26
+ 65	+ 29	+ 50	+ 67	+ 90	+ 87

E

85	37	22	24	92	64
+ 32	+ 15	+ 45	+ 15	+ 47	+ 46

F

79	87	81	60	92	66
+ 24	+ 47	+ 53	+ 89	+ 29	+ 56

G

37	92	42	34	67	74
+ 95	+ 71	+ 82	+ 44	+ 50	+ 50

H

20	82	61	13	37	15
+ 82	+ 50	+ 47	+ 94	+ 43	+ 53

I

85	32	13	50	45	24
+ 13	+ 63	+ 13	+ 87	+ 56	+ 23

J

56	44	95	84	13	38
+ 48	+ 44	+ 48	+ 66	+ 62	+ 80

A
43	94	53	35	78	71
+ 75	+ 80	+ 56	+ 67	+ 68	+ 51

B
69	91	55	66	68	81
+ 51	+ 66	+ 90	+ 15	+ 81	+ 85

C
16	45	98	21	55	87
+ 34	+ 36	+ 68	+ 17	+ 37	+ 62

D
99	33	60	14	62	70
+ 86	+ 48	+ 79	+ 42	+ 86	+ 72

E
30	85	60	37	61	14
+ 62	+ 26	+ 89	+ 88	+ 36	+ 86

F
41	22	98	57	21	93
+ 27	+ 98	+ 16	+ 25	+ 33	+ 58

G
53	61	93	21	70	97
+ 64	+ 18	+ 57	+ 61	+ 73	+ 16

H
52	99	28	41	17	83
+ 64	+ 32	+ 53	+ 52	+ 40	+ 40

I
39	91	96	45	59	81
+ 79	+ 29	+ 21	+ 49	+ 35	+ 89

J
63	13	38	65	85	68
+ 71	+ 65	+ 80	+ 56	+ 83	+ 17

A
| 69 | 16 | 48 | 59 | 63 | 88 |
| + 29 | + 92 | + 19 | + 83 | + 80 | + 68 |

B
| 73 | 75 | 62 | 47 | 63 | 48 |
| + 26 | + 49 | + 40 | + 50 | + 32 | + 72 |

C
| 70 | 53 | 81 | 89 | 24 | 90 |
| + 81 | + 48 | + 77 | + 19 | + 69 | + 95 |

D
| 96 | 77 | 26 | 42 | 49 | 96 |
| + 80 | + 68 | + 39 | + 69 | + 61 | + 99 |

E
| 10 | 62 | 93 | 54 | 50 | 31 |
| + 19 | + 48 | + 72 | + 82 | + 26 | + 20 |

30

F
| 71 | 37 | 44 | 21 | 14 | 81 |
| + 52 | + 74 | + 32 | + 17 | + 14 | + 47 |

G
| 87 | 25 | 28 | 94 | 26 | 37 |
| + 74 | + 76 | + 17 | + 76 | + 87 | + 78 |

H
| 34 | 67 | 73 | 56 | 45 | 12 |
| + 30 | + 41 | + 22 | + 18 | + 26 | + 54 |

I
| 31 | 41 | 15 | 83 | 30 | 29 |
| + 14 | + 74 | + 23 | + 25 | + 48 | + 42 |

J
| 60 | 75 | 12 | 43 | 49 | 49 |
| + 95 | + 45 | + 36 | + 59 | + 96 | + 36 |

60

Day: 13

Date:

Score: /60

Name:

Time: :

Rating: ☆☆☆☆☆

A	58 + 25	58 + 37	76 + 42	35 + 25	52 + 46	21 + 31
B	58 + 57	94 + 29	36 + 40	93 + 85	15 + 35	54 + 15
C	57 + 61	80 + 75	23 + 69	78 + 90	81 + 88	34 + 73
D	54 + 42	17 + 67	93 + 22	45 + 37	39 + 96	16 + 79
E	32 + 49	33 + 22	76 + 87	86 + 70	66 + 78	13 + 21
F	41 + 28	58 + 90	22 + 14	97 + 36	78 + 98	72 + 16
G	41 + 31	85 + 39	65 + 19	18 + 39	40 + 57	87 + 56
H	52 + 49	42 + 25	67 + 92	89 + 48	24 + 67	50 + 26
I	47 + 62	69 + 78	23 + 62	10 + 52	63 + 12	49 + 68
J	72 + 39	35 + 57	52 + 44	55 + 92	90 + 50	81 + 71

30

60

A	49 + 93	97 + 94	97 + 47	76 + 89	74 + 44	60 + 50
B	81 + 96	75 + 35	51 + 52	67 + 40	12 + 28	29 + 96
C	89 + 17	99 + 84	19 + 52	30 + 13	35 + 97	88 + 94
D	16 + 72	13 + 36	69 + 81	32 + 62	90 + 84	85 + 27
E	96 + 79	26 + 64	35 + 44	44 + 92	54 + 27	47 + 89
F	68 + 56	84 + 49	89 + 58	29 + 18	97 + 13	40 + 10
G	15 + 74	58 + 38	34 + 28	72 + 41	97 + 90	48 + 46
H	94 + 52	13 + 63	48 + 39	11 + 20	98 + 77	38 + 65
I	94 + 48	96 + 63	11 + 10	16 + 22	47 + 12	47 + 27
J	51 + 89	31 + 14	95 + 99	32 + 31	75 + 16	95 + 75

A

29	26	85	21	55	78
+ 56	+ 96	+ 79	+ 99	+ 75	+ 32

B

59	26	88	75	34	30
+ 28	+ 88	+ 68	+ 30	+ 97	+ 66

C

48	88	35	48	66	49
+ 95	+ 60	+ 25	+ 62	+ 17	+ 54

D

46	15	54	39	62	26
+ 52	+ 23	+ 99	+ 28	+ 59	+ 42

E

48	44	88	69	51	80
+ 80	+ 82	+ 53	+ 75	+ 81	+ 32

(30)

F

77	54	18	17	82	52
+ 84	+ 97	+ 66	+ 89	+ 17	+ 17

G

33	26	24	34	55	84
+ 41	+ 56	+ 14	+ 63	+ 19	+ 35

H

36	77	39	72	92	75
+ 78	+ 61	+ 67	+ 87	+ 42	+ 56

I

75	31	70	49	92	49
+ 56	+ 43	+ 86	+ 39	+ 40	+ 97

J

93	24	25	29	99	28
+ 99	+ 12	+ 81	+ 42	+ 97	+ 46

(60)

A

23	87	63	92	60	24
+ 15	+ 17	+ 60	+ 34	+ 24	+ 94

B

23	86	50	14	99	35
+ 24	+ 77	+ 84	+ 72	+ 81	+ 10

C

85	93	79	96	38	82
+ 42	+ 31	+ 53	+ 78	+ 13	+ 99

D

52	58	38	60	56	66
+ 75	+ 70	+ 25	+ 53	+ 36	+ 90

E

82	67	78	27	72	26
+ 85	+ 42	+ 13	+ 68	+ 18	+ 88

F

75	48	47	86	95	37
+ 23	+ 90	+ 92	+ 94	+ 15	+ 31

G

64	52	13	37	23	39
+ 63	+ 58	+ 53	+ 20	+ 83	+ 48

H

73	98	14	34	15	85
+ 75	+ 83	+ 48	+ 20	+ 96	+ 66

I

58	65	29	97	19	34
+ 64	+ 72	+ 75	+ 62	+ 63	+ 34

J

17	97	25	30	49	30
+ 22	+ 56	+ 70	+ 41	+ 51	+ 99

A
39	50	53	65	30	97
+ 85	+ 56	+ 73	+ 91	+ 65	+ 76

B
73	64	95	32	99	42
+ 64	+ 42	+ 39	+ 31	+ 73	+ 68

C
94	91	28	51	60	30
+ 74	+ 56	+ 81	+ 24	+ 44	+ 85

D
65	87	51	20	34	30
+ 71	+ 45	+ 48	+ 62	+ 68	+ 47

E
23	60	41	53	17	66
+ 37	+ 87	+ 42	+ 60	+ 90	+ 49

F
55	67	73	22	51	77
+ 39	+ 58	+ 47	+ 94	+ 81	+ 14

G
17	59	20	28	24	25
+ 17	+ 83	+ 13	+ 37	+ 97	+ 33

H
54	77	59	35	75	97
+ 99	+ 33	+ 97	+ 14	+ 75	+ 28

I
61	40	77	56	63	70
+ 47	+ 16	+ 13	+ 65	+ 95	+ 61

J
86	80	33	84	67	78
+ 30	+ 21	+ 17	+ 20	+ 68	+ 96

A
| 92
 + 69 | 22
 + 68 | 90
 + 43 | 36
 + 56 | 47
 + 93 | 37
 + 70 |

B
| 69
 + 68 | 76
 + 79 | 76
 + 81 | 28
 + 16 | 76
 + 12 | 17
 + 81 |

C
| 13
 + 70 | 69
 + 38 | 42
 + 32 | 99
 + 31 | 85
 + 46 | 78
 + 94 |

D
| 17
 + 77 | 87
 + 45 | 80
 + 21 | 72
 + 85 | 48
 + 60 | 87
 + 92 |

E
| 69
 + 47 | 53
 + 28 | 26
 + 14 | 79
 + 52 | 64
 + 49 | 62
 + 78 |

F
| 34
 + 74 | 71
 + 48 | 49
 + 75 | 41
 + 72 | 21
 + 91 | 37
 + 97 |

G
| 81
 + 21 | 34
 + 13 | 96
 + 90 | 36
 + 61 | 76
 + 55 | 96
 + 43 |

H
| 27
 + 57 | 13
 + 85 | 64
 + 39 | 65
 + 21 | 25
 + 72 | 80
 + 71 |

I
| 42
 + 11 | 11
 + 51 | 43
 + 75 | 28
 + 61 | 16
 + 89 | 49
 + 50 |

J
| 33
 + 82 | 87
 + 94 | 26
 + 48 | 87
 + 59 | 81
 + 25 | 12
 + 41 |

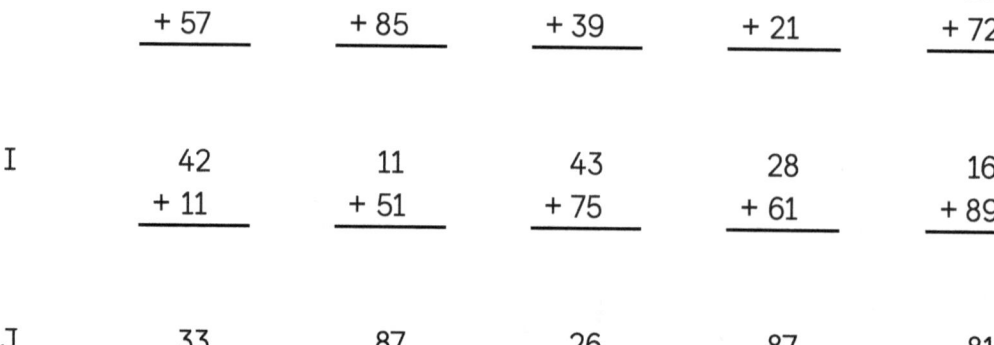

A	39 + 19	13 + 63	22 + 55	61 + 77	76 + 38	13 + 94
B	78 + 40	43 + 51	22 + 33	34 + 71	79 + 84	71 + 25
C	36 + 77	27 + 43	81 + 10	63 + 20	86 + 89	96 + 40
D	64 + 44	70 + 95	33 + 33	70 + 18	73 + 38	29 + 35
E	69 + 76	42 + 87	30 + 55	21 + 99	98 + 24	28 + 92
F	35 + 56	53 + 25	95 + 40	42 + 53	55 + 18	95 + 15
G	75 + 49	87 + 62	60 + 38	47 + 72	19 + 54	93 + 66
H	74 + 39	42 + 68	80 + 47	29 + 15	95 + 85	48 + 70
I	42 + 30	89 + 87	64 + 62	17 + 67	54 + 41	44 + 23
J	66 + 87	11 + 96	75 + 63	59 + 53	90 + 49	24 + 86

A
| 31 | 35 | 67 | 27 | 51 | 96 |
| + 65 | + 88 | + 90 | + 62 | + 43 | + 64 |

B
| 50 | 52 | 24 | 87 | 76 | 74 |
| + 19 | + 72 | + 84 | + 49 | + 62 | + 28 |

C
| 51 | 53 | 19 | 36 | 43 | 94 |
| + 87 | + 41 | + 49 | + 36 | + 61 | + 75 |

D
| 38 | 95 | 40 | 71 | 98 | 95 |
| + 90 | + 63 | + 23 | + 20 | + 32 | + 59 |

E
| 62 | 70 | 24 | 40 | 31 | 62 |
| + 20 | + 44 | + 97 | + 58 | + 91 | + 76 |

F
| 23 | 68 | 97 | 97 | 36 | 77 |
| + 57 | + 78 | + 57 | + 76 | + 17 | + 56 |

G
| 83 | 58 | 74 | 33 | 82 | 24 |
| + 11 | + 66 | + 62 | + 55 | + 17 | + 52 |

H
| 81 | 51 | 15 | 44 | 30 | 95 |
| + 98 | + 41 | + 35 | + 28 | + 20 | + 58 |

I
| 30 | 39 | 97 | 70 | 27 | 14 |
| + 49 | + 83 | + 63 | + 14 | + 39 | + 97 |

J
| 66 | 17 | 71 | 18 | 52 | 26 |
| + 31 | + 53 | + 42 | + 85 | + 97 | + 37 |

A
| 42 | 20 | 32 | 38 | 54 | 95 |
| + 69 | + 43 | + 19 | + 51 | + 22 | + 22 |

B
| 90 | 86 | 91 | 78 | 56 | 43 |
| + 73 | + 41 | + 47 | + 71 | + 92 | + 28 |

C
| 20 | 32 | 50 | 57 | 30 | 45 |
| + 12 | + 65 | + 21 | + 57 | + 44 | + 17 |

D
| 11 | 92 | 38 | 17 | 98 | 92 |
| + 57 | + 49 | + 92 | + 88 | + 66 | + 34 |

E
| 24 | 19 | 18 | 46 | 40 | 14 |
| + 84 | + 13 | + 54 | + 11 | + 84 | + 13 |

F
| 38 | 39 | 10 | 39 | 14 | 55 |
| + 66 | + 30 | + 34 | + 62 | + 94 | + 45 |

G
| 39 | 24 | 87 | 99 | 59 | 37 |
| + 36 | + 76 | + 72 | + 66 | + 29 | + 44 |

H
| 36 | 23 | 30 | 57 | 95 | 93 |
| + 33 | + 64 | + 86 | + 86 | + 33 | + 82 |

I
| 78 | 72 | 66 | 11 | 80 | 70 |
| + 21 | + 43 | + 19 | + 80 | + 40 | + 18 |

J
| 27 | 68 | 54 | 80 | 86 | 24 |
| + 50 | + 95 | + 36 | + 68 | + 19 | + 88 |

A

30	98	77	62	87	90
+ 26	+ 45	+ 14	+ 54	+ 27	+ 91

B

50	77	80	87	69	11
+ 29	+ 99	+ 56	+ 99	+ 98	+ 22

C

82	62	10	27	60	12
+ 93	+ 49	+ 29	+ 88	+ 42	+ 16

D

71	81	13	84	15	89
+ 99	+ 34	+ 76	+ 19	+ 90	+ 98

E

60	26	92	14	77	73
+ 17	+ 92	+ 90	+ 53	+ 47	+ 59

F

91	83	85	43	85	44
+ 43	+ 87	+ 67	+ 79	+ 24	+ 80

G

19	70	71	41	30	53
+ 38	+ 10	+ 29	+ 46	+ 80	+ 10

H

29	83	50	35	16	53
+ 86	+ 61	+ 35	+ 14	+ 63	+ 61

I

28	38	70	74	81	60
+ 37	+ 85	+ 52	+ 93	+ 78	+ 47

J

83	96	83	84	42	25
+ 18	+ 79	+ 86	+ 20	+ 96	+ 66

A

| 60
+ 46 | 70
+ 21 | 89
+ 58 | 89
+ 53 | 21
+ 14 | 29
+ 10 |

B

| 61
+ 55 | 55
+ 41 | 62
+ 42 | 69
+ 32 | 99
+ 17 | 89
+ 43 |

C

| 91
+ 29 | 97
+ 13 | 15
+ 42 | 75
+ 94 | 15
+ 87 | 34
+ 78 |

D

| 56
+ 93 | 70
+ 68 | 70
+ 79 | 23
+ 69 | 83
+ 71 | 82
+ 38 |

E

| 69
+ 34 | 67
+ 16 | 56
+ 18 | 65
+ 37 | 28
+ 57 | 95
+ 74 |

F

| 94
+ 19 | 75
+ 49 | 80
+ 95 | 41
+ 48 | 37
+ 60 | 33
+ 16 |

G

| 29
+ 63 | 61
+ 96 | 82
+ 14 | 86
+ 46 | 68
+ 92 | 11
+ 50 |

H

| 23
+ 57 | 53
+ 30 | 79
+ 16 | 76
+ 64 | 91
+ 76 | 42
+ 46 |

I

| 77
+ 44 | 70
+ 67 | 10
+ 61 | 80
+ 82 | 66
+ 92 | 96
+ 94 |

J

| 78
+ 41 | 77
+ 33 | 77
+ 60 | 34
+ 53 | 51
+ 97 | 14
+ 51 |

Day:	24		Date:		Score:	/60
Name:			Time:	:	Rating:	☆☆☆☆☆

A	54 + 62	95 + 90	29 + 64	55 + 23	93 + 30	14 + 79
B	33 + 57	91 + 61	24 + 26	88 + 13	79 + 95	30 + 20
C	62 + 67	74 + 13	68 + 70	36 + 29	34 + 44	55 + 90
D	66 + 54	84 + 51	54 + 98	88 + 28	27 + 85	12 + 51
E	36 + 50	26 + 65	26 + 85	27 + 40	65 + 32	72 + 23
F	44 + 96	37 + 17	33 + 90	86 + 21	21 + 18	67 + 94
G	46 + 12	66 + 96	90 + 46	86 + 85	64 + 82	36 + 83
H	48 + 80	64 + 95	82 + 45	66 + 87	70 + 70	47 + 79
I	15 + 26	82 + 69	85 + 93	73 + 30	14 + 12	88 + 60
J	22 + 70	63 + 41	23 + 80	82 + 47	10 + 82	31 + 12

A
| 23 | 86 | 77 | 41 | 46 | 97 |
| + 65 | + 39 | + 19 | + 89 | + 74 | + 52 |

B
| 27 | 79 | 21 | 50 | 22 | 87 |
| + 61 | + 19 | + 38 | + 56 | + 67 | + 12 |

C
| 28 | 78 | 62 | 11 | 47 | 56 |
| + 98 | + 26 | + 20 | + 11 | + 38 | + 91 |

D
| 50 | 95 | 21 | 31 | 60 | 70 |
| + 22 | + 54 | + 79 | + 12 | + 87 | + 64 |

E
| 47 | 21 | 27 | 20 | 87 | 77 |
| + 33 | + 46 | + 35 | + 90 | + 16 | + 17 |

F
| 40 | 51 | 73 | 66 | 88 | 99 |
| + 47 | + 72 | + 64 | + 69 | + 90 | + 27 |

G
| 84 | 50 | 90 | 69 | 75 | 95 |
| + 81 | + 64 | + 85 | + 24 | + 65 | + 33 |

H
| 27 | 38 | 86 | 12 | 16 | 46 |
| + 60 | + 49 | + 66 | + 79 | + 53 | + 86 |

I
| 64 | 89 | 68 | 37 | 41 | 59 |
| + 38 | + 94 | + 57 | + 50 | + 75 | + 56 |

J
| 38 | 36 | 81 | 32 | 66 | 96 |
| + 17 | + 53 | + 43 | + 19 | + 42 | + 62 |

A
13	11	51	42	26	21
+ 49	+ 96	+ 99	+ 84	+ 66	+ 33

B
36	40	55	11	52	75
+ 93	+ 44	+ 32	+ 27	+ 77	+ 66

C
58	18	40	40	47	41
+ 49	+ 94	+ 34	+ 69	+ 57	+ 18

D
39	21	35	39	60	16
+ 92	+ 93	+ 55	+ 99	+ 48	+ 46

E
68	33	61	21	58	41
+ 34	+ 86	+ 83	+ 96	+ 21	+ 51

F
33	35	38	28	86	65
+ 76	+ 60	+ 69	+ 96	+ 11	+ 77

G
76	42	41	49	41	39
+ 22	+ 20	+ 92	+ 99	+ 32	+ 64

H
16	79	46	16	91	28
+ 55	+ 57	+ 59	+ 98	+ 84	+ 52

I
55	45	96	41	92	64
+ 19	+ 75	+ 75	+ 44	+ 97	+ 61

J
58	45	39	17	48	96
+ 55	+ 34	+ 79	+ 80	+ 75	+ 64

A
39	54	69	97	26	57
+ 33	+ 29	+ 33	+ 66	+ 20	+ 94

B
31	71	81	68	73	64
+ 57	+ 90	+ 65	+ 12	+ 89	+ 26

C
76	72	62	24	42	65
+ 24	+ 48	+ 39	+ 26	+ 75	+ 42

D
81	66	83	93	77	64
+ 18	+ 72	+ 52	+ 97	+ 12	+ 23

E
51	49	55	21	23	32
+ 60	+ 20	+ 71	+ 25	+ 72	+ 41

F
39	60	55	54	99	99
+ 58	+ 35	+ 66	+ 25	+ 28	+ 42

G
62	47	91	45	61	86
+ 48	+ 41	+ 65	+ 77	+ 78	+ 12

H
32	40	18	24	15	10
+ 85	+ 25	+ 87	+ 13	+ 58	+ 10

I
67	12	18	95	13	70
+ 91	+ 19	+ 80	+ 39	+ 31	+ 61

J
20	13	53	82	96	73
+ 37	+ 97	+ 36	+ 80	+ 49	+ 40

A

42	50	45	46	77	14
+ 89	+ 39	+ 38	+ 14	+ 74	+ 11

B

62	57	88	14	18	15
+ 73	+ 40	+ 46	+ 23	+ 45	+ 40

C

23	54	70	64	32	86
+ 26	+ 28	+ 17	+ 86	+ 60	+ 95

D

61	77	18	80	89	42
+ 26	+ 70	+ 49	+ 61	+ 17	+ 84

E

49	30	38	72	56	40
+ 87	+ 65	+ 42	+ 78	+ 77	+ 74

F

24	63	31	50	84	44
+ 65	+ 44	+ 21	+ 97	+ 88	+ 67

G

82	50	38	26	47	10
+ 95	+ 39	+ 57	+ 72	+ 53	+ 98

H

68	59	47	77	94	26
+ 48	+ 97	+ 10	+ 22	+ 34	+ 38

I

43	50	22	39	63	39
+ 42	+ 81	+ 35	+ 48	+ 59	+ 18

J

14	90	74	69	86	77
+ 63	+ 89	+ 72	+ 13	+ 34	+ 90

A
$$27 + 35 \qquad 16 + 44 \qquad 23 + 90 \qquad 63 + 79 \qquad 76 + 40 \qquad 80 + 40$$

B
$$65 + 45 \qquad 93 + 29 \qquad 11 + 40 \qquad 17 + 37 \qquad 42 + 29 \qquad 87 + 58$$

C
$$85 + 55 \qquad 25 + 17 \qquad 45 + 64 \qquad 12 + 63 \qquad 45 + 60 \qquad 47 + 57$$

D
$$23 + 99 \qquad 63 + 28 \qquad 86 + 51 \qquad 54 + 86 \qquad 24 + 48 \qquad 98 + 46$$

E
$$73 + 37 \qquad 12 + 34 \qquad 28 + 78 \qquad 32 + 98 \qquad 84 + 91 \qquad 10 + 46$$

F
$$53 + 27 \qquad 63 + 96 \qquad 36 + 82 \qquad 33 + 98 \qquad 10 + 36 \qquad 21 + 97$$

G
$$30 + 94 \qquad 79 + 22 \qquad 67 + 32 \qquad 26 + 44 \qquad 79 + 97 \qquad 66 + 25$$

H
$$99 + 31 \qquad 57 + 80 \qquad 44 + 23 \qquad 17 + 97 \qquad 68 + 34 \qquad 22 + 38$$

I
$$93 + 98 \qquad 11 + 86 \qquad 64 + 96 \qquad 82 + 28 \qquad 56 + 71 \qquad 60 + 77$$

J
$$44 + 39 \qquad 37 + 69 \qquad 69 + 64 \qquad 44 + 41 \qquad 73 + 29 \qquad 85 + 68$$

A	77 + 89	92 + 28	68 + 98	47 + 95	80 + 86	40 + 29
B	73 + 60	64 + 57	12 + 80	16 + 45	44 + 74	30 + 96
C	69 + 81	34 + 54	82 + 65	29 + 18	78 + 21	57 + 72
D	48 + 84	23 + 44	92 + 90	33 + 24	38 + 58	46 + 21
E	76 + 29	68 + 84	36 + 17	34 + 83	71 + 48	46 + 43
F	26 + 12	82 + 22	32 + 35	36 + 64	54 + 79	63 + 33
G	70 + 95	47 + 42	43 + 45	59 + 31	54 + 57	82 + 47
H	20 + 32	36 + 77	83 + 49	71 + 11	43 + 85	71 + 65
I	21 + 93	76 + 76	36 + 93	54 + 42	23 + 11	21 + 82
J	63 + 52	84 + 80	44 + 80	24 + 76	14 + 48	92 + 93

30

60

A
| 72 | 42 | 44 | 27 | 69 | 73 |
| + 39 | + 38 | + 25 | + 87 | + 41 | + 42 |

B
| 24 | 36 | 29 | 77 | 33 | 10 |
| + 25 | + 95 | + 43 | + 79 | + 54 | + 15 |

C
| 24 | 37 | 44 | 81 | 55 | 95 |
| + 87 | + 77 | + 42 | + 49 | + 27 | + 92 |

D
| 22 | 28 | 20 | 21 | 84 | 46 |
| + 92 | + 27 | + 88 | + 64 | + 79 | + 78 |

E
| 36 | 56 | 44 | 40 | 53 | 47 |
| + 97 | + 60 | + 69 | + 12 | + 47 | + 11 |

F
| 83 | 79 | 82 | 86 | 45 | 62 |
| + 11 | + 51 | + 45 | + 11 | + 27 | + 81 |

G
| 95 | 35 | 86 | 55 | 68 | 82 |
| + 49 | + 85 | + 32 | + 48 | + 56 | + 39 |

H
| 44 | 22 | 92 | 12 | 67 | 57 |
| + 25 | + 47 | + 53 | + 17 | + 63 | + 77 |

I
| 63 | 61 | 55 | 42 | 40 | 86 |
| + 19 | + 15 | + 43 | + 53 | + 72 | + 25 |

J
| 77 | 30 | 11 | 32 | 80 | 40 |
| + 25 | + 16 | + 46 | + 31 | + 38 | + 78 |

Day:	32			**Date:**		**Score:** /60
Name:				**Time:** :		**Rating:** ☆☆☆☆☆

A	47 + 56	27 + 63	61 + 33	67 + 34	96 + 65	10 + 96
B	69 + 86	99 + 34	27 + 62	38 + 24	80 + 37	52 + 17
C	32 + 12	61 + 15	60 + 90	58 + 37	96 + 26	45 + 74
D	48 + 59	35 + 14	52 + 84	60 + 80	26 + 95	49 + 58
E	75 + 56	15 + 59	38 + 59	19 + 60	79 + 96	45 + 84
F	31 + 13	85 + 41	49 + 79	53 + 40	88 + 79	19 + 96
G	65 + 72	50 + 11	37 + 79	15 + 66	17 + 90	76 + 77
H	16 + 64	41 + 57	27 + 14	36 + 54	94 + 52	16 + 11
I	40 + 26	60 + 49	10 + 18	86 + 97	92 + 54	98 + 72
J	27 + 60	67 + 59	63 + 83	78 + 86	95 + 51	82 + 63

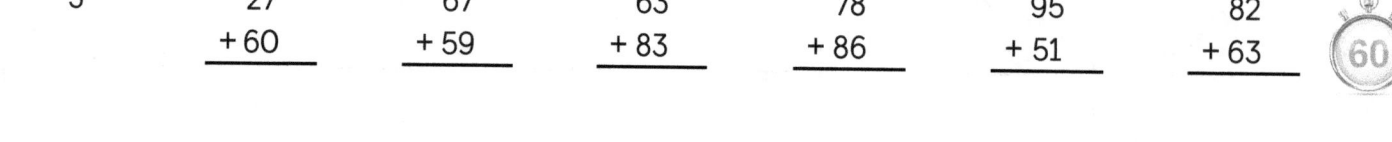

A	78 + 98	60 + 84	96 + 59	89 + 55	37 + 26	93 + 96
B	58 + 66	32 + 27	66 + 59	15 + 24	83 + 52	93 + 12
C	47 + 34	17 + 99	49 + 94	43 + 88	74 + 21	91 + 20
D	59 + 77	57 + 65	43 + 63	37 + 71	25 + 68	30 + 54
E	61 + 44	29 + 51	61 + 54	29 + 53	49 + 33	99 + 12
F	54 + 33	52 + 75	84 + 24	21 + 55	52 + 78	59 + 21
G	94 + 21	33 + 34	93 + 52	66 + 38	92 + 20	11 + 45
H	90 + 88	13 + 99	43 + 43	71 + 61	29 + 35	74 + 64
I	67 + 75	44 + 42	46 + 95	13 + 51	58 + 81	16 + 69
J	59 + 85	59 + 50	62 + 31	93 + 49	62 + 52	97 + 64

30

60

A
$$65 + 96 \quad 45 + 19 \quad 61 + 94 \quad 53 + 48 \quad 30 + 76 \quad 15 + 21$$

B
$$39 + 27 \quad 89 + 76 \quad 86 + 19 \quad 99 + 59 \quad 19 + 43 \quad 91 + 26$$

C
$$11 + 76 \quad 44 + 22 \quad 36 + 42 \quad 68 + 69 \quad 39 + 96 \quad 59 + 27$$

D
$$83 + 21 \quad 16 + 91 \quad 44 + 81 \quad 60 + 47 \quad 87 + 14 \quad 73 + 38$$

E
$$20 + 72 \quad 58 + 15 \quad 97 + 43 \quad 75 + 14 \quad 88 + 17 \quad 81 + 49$$

F
$$11 + 31 \quad 24 + 35 \quad 11 + 31 \quad 60 + 24 \quad 46 + 59 \quad 25 + 36$$

G
$$66 + 88 \quad 44 + 57 \quad 50 + 21 \quad 16 + 59 \quad 43 + 71 \quad 79 + 79$$

H
$$49 + 49 \quad 11 + 26 \quad 51 + 80 \quad 14 + 56 \quad 46 + 19 \quad 35 + 36$$

I
$$48 + 13 \quad 97 + 24 \quad 98 + 29 \quad 54 + 36 \quad 92 + 11 \quad 42 + 52$$

J
$$86 + 44 \quad 13 + 67 \quad 62 + 87 \quad 49 + 96 \quad 90 + 81 \quad 71 + 82$$

A	88 + 20	38 + 74	68 + 53	26 + 21	67 + 10	24 + 90
B	84 + 67	95 + 88	33 + 36	59 + 54	68 + 55	43 + 63
C	35 + 96	47 + 49	92 + 32	53 + 58	95 + 52	52 + 97
D	81 + 70	81 + 24	65 + 84	86 + 64	83 + 11	42 + 95
E	70 + 93	52 + 14	71 + 15	42 + 12	11 + 11	31 + 34
F	30 + 97	59 + 95	78 + 77	45 + 95	64 + 59	22 + 50
G	80 + 37	81 + 17	25 + 96	59 + 34	35 + 70	27 + 42
H	59 + 12	60 + 67	93 + 78	72 + 98	49 + 53	14 + 70
I	87 + 15	14 + 92	52 + 68	39 + 85	65 + 26	21 + 73
J	76 + 58	12 + 11	42 + 50	52 + 16	92 + 95	78 + 17

A
$$45 + 11 \quad 64 + 65 \quad 19 + 73 \quad 93 + 69 \quad 18 + 70 \quad 58 + 14$$

B
$$56 + 45 \quad 37 + 25 \quad 48 + 13 \quad 41 + 55 \quad 95 + 68 \quad 33 + 32$$

C
$$90 + 44 \quad 43 + 27 \quad 21 + 99 \quad 28 + 27 \quad 60 + 62 \quad 34 + 95$$

D
$$34 + 53 \quad 63 + 18 \quad 32 + 54 \quad 45 + 18 \quad 72 + 39 \quad 67 + 44$$

E
$$33 + 65 \quad 43 + 40 \quad 68 + 93 \quad 55 + 74 \quad 92 + 13 \quad 65 + 44$$

F
$$89 + 86 \quad 56 + 99 \quad 97 + 31 \quad 21 + 13 \quad 12 + 44 \quad 45 + 70$$

G
$$78 + 72 \quad 26 + 90 \quad 38 + 72 \quad 86 + 85 \quad 56 + 87 \quad 58 + 63$$

H
$$41 + 37 \quad 21 + 55 \quad 64 + 16 \quad 71 + 18 \quad 50 + 98 \quad 15 + 11$$

I
$$45 + 33 \quad 95 + 99 \quad 19 + 34 \quad 91 + 76 \quad 87 + 79 \quad 53 + 57$$

J
$$82 + 87 \quad 10 + 34 \quad 98 + 62 \quad 79 + 99 \quad 52 + 23 \quad 26 + 26$$

A

| 34
+ 15 | 25
+ 51 | 65
+ 31 | 54
+ 66 | 34
+ 20 | 65
+ 25 |

B

| 67
+ 13 | 53
+ 60 | 13
+ 74 | 97
+ 26 | 54
+ 63 | 62
+ 41 |

C

| 97
+ 78 | 81
+ 42 | 99
+ 84 | 22
+ 91 | 63
+ 54 | 98
+ 16 |

D

| 53
+ 43 | 76
+ 79 | 49
+ 92 | 69
+ 38 | 38
+ 36 | 47
+ 11 |

E

| 71
+ 63 | 93
+ 78 | 33
+ 68 | 65
+ 54 | 70
+ 13 | 56
+ 72 |

F

| 58
+ 85 | 70
+ 87 | 64
+ 82 | 67
+ 49 | 88
+ 78 | 44
+ 53 |

G

| 35
+ 81 | 97
+ 32 | 46
+ 75 | 90
+ 87 | 55
+ 81 | 71
+ 96 |

H

| 93
+ 48 | 60
+ 97 | 79
+ 28 | 87
+ 58 | 82
+ 25 | 82
+ 42 |

I

| 96
+ 59 | 47
+ 82 | 85
+ 99 | 13
+ 54 | 17
+ 80 | 87
+ 24 |

J

| 50
+ 55 | 58
+ 62 | 32
+ 75 | 65
+ 58 | 93
+ 24 | 75
+ 66 |

A

96	44	81	45	52	71
+ 14	+ 26	+ 38	+ 41	+ 41	+ 59

B

65	79	17	75	42	28
+ 15	+ 42	+ 23	+ 79	+ 36	+ 43

C

29	37	37	65	85	57
+ 72	+ 23	+ 37	+ 66	+ 90	+ 35

D

55	36	13	35	15	28
+ 86	+ 84	+ 93	+ 58	+ 27	+ 24

E

19	88	36	65	58	98
+ 50	+ 18	+ 72	+ 43	+ 93	+ 55

F

62	64	40	55	64	23
+ 86	+ 13	+ 98	+ 49	+ 53	+ 59

G

99	91	98	74	67	91
+ 17	+ 56	+ 68	+ 19	+ 69	+ 29

H

77	69	44	47	30	33
+ 63	+ 43	+ 52	+ 40	+ 65	+ 30

I

25	14	76	22	45	71
+ 75	+ 36	+ 76	+ 71	+ 43	+ 95

J

24	51	62	41	46	20
+ 30	+ 93	+ 72	+ 60	+ 33	+ 19

A	85 + 96	61 + 14	45 + 36	92 + 62	97 + 31	13 + 30
B	97 + 56	63 + 24	93 + 59	45 + 32	70 + 42	35 + 67
C	12 + 43	74 + 11	36 + 24	81 + 79	48 + 72	65 + 61
D	70 + 37	65 + 56	80 + 30	86 + 36	59 + 30	93 + 74
E	10 + 52	48 + 64	19 + 23	20 + 25	37 + 17	19 + 57
F	23 + 48	94 + 63	60 + 90	97 + 17	10 + 14	79 + 12
G	91 + 98	48 + 69	33 + 45	47 + 72	64 + 51	89 + 96
H	26 + 88	58 + 57	60 + 43	90 + 44	32 + 17	73 + 42
I	35 + 26	92 + 83	11 + 37	20 + 38	93 + 80	80 + 90
J	40 + 94	66 + 30	71 + 19	26 + 83	99 + 75	26 + 50

A	85 + 35	88 + 78	82 + 21	52 + 40	65 + 58	81 + 40
B	59 + 38	93 + 87	97 + 25	61 + 91	40 + 20	79 + 97
C	99 + 39	44 + 75	76 + 48	71 + 85	75 + 60	51 + 38
D	30 + 23	51 + 20	46 + 96	26 + 57	25 + 93	17 + 88
E	72 + 38	71 + 78	43 + 19	20 + 15	61 + 37	92 + 94
F	34 + 87	51 + 56	13 + 99	13 + 48	90 + 80	87 + 23
G	56 + 25	17 + 44	76 + 26	16 + 88	36 + 90	26 + 88
H	24 + 67	43 + 85	92 + 70	45 + 86	93 + 63	24 + 26
I	99 + 48	79 + 47	40 + 72	59 + 71	59 + 44	14 + 29
J	69 + 30	26 + 49	54 + 70	19 + 26	50 + 91	66 + 96

30

60

A
| 66
+ 32 | 16
+ 89 | 65
+ 35 | 19
+ 42 | 49
+ 92 | 36
+ 59 |

B
| 92
+ 80 | 30
+ 11 | 44
+ 42 | 77
+ 84 | 59
+ 93 | 95
+ 39 |

C
| 22
+ 21 | 52
+ 84 | 63
+ 42 | 81
+ 14 | 90
+ 31 | 47
+ 96 |

D
| 21
+ 83 | 50
+ 21 | 11
+ 84 | 99
+ 14 | 52
+ 71 | 86
+ 61 |

E
| 17
+ 32 | 57
+ 28 | 76
+ 51 | 54
+ 88 | 19
+ 87 | 54
+ 56 |

F
| 69
+ 90 | 40
+ 60 | 12
+ 63 | 96
+ 13 | 10
+ 36 | 29
+ 56 |

G
| 77
+ 52 | 69
+ 55 | 17
+ 16 | 51
+ 96 | 58
+ 92 | 21
+ 10 |

H
| 65
+ 73 | 52
+ 80 | 53
+ 96 | 76
+ 93 | 88
+ 20 | 94
+ 73 |

I
| 38
+ 66 | 55
+ 82 | 41
+ 62 | 25
+ 54 | 95
+ 77 | 46
+ 12 |

J
| 74
+ 50 | 26
+ 60 | 60
+ 75 | 29
+ 35 | 28
+ 34 | 75
+ 86 |

A
95 + 14   75 + 18   57 + 79   30 + 68   71 + 33   44 + 74

B
41 + 64   99 + 91   58 + 98   89 + 13   36 + 50   67 + 99

C
25 + 36   60 + 90   58 + 87   21 + 57   16 + 81   65 + 47

D
39 + 77   98 + 44   74 + 14   52 + 43   36 + 83   19 + 68

E
18 + 75   97 + 30   41 + 77   67 + 88   37 + 94   77 + 51

F
95 + 37   90 + 52   55 + 58   86 + 71   65 + 50   46 + 56

G
90 + 91   92 + 35   57 + 17   59 + 52   84 + 21   50 + 97

H
77 + 57   65 + 61   46 + 42   86 + 64   50 + 78   29 + 88

I
54 + 98   81 + 82   10 + 78   78 + 12   10 + 36   31 + 48

J
13 + 69   65 + 30   96 + 61   88 + 13   35 + 38   10 + 72

A
| 86 | 39 | 12 | 69 | 38 | 22 |
| + 59 | + 97 | + 55 | + 70 | + 11 | + 61 |

B
| 74 | 29 | 52 | 31 | 99 | 48 |
| + 20 | + 78 | + 26 | + 11 | + 44 | + 23 |

C
| 83 | 46 | 89 | 96 | 22 | 33 |
| + 64 | + 16 | + 87 | + 99 | + 85 | + 17 |

D
| 31 | 51 | 78 | 74 | 31 | 98 |
| + 92 | + 88 | + 64 | + 31 | + 96 | + 63 |

E
| 31 | 81 | 87 | 19 | 96 | 16 |
| + 81 | + 50 | + 76 | + 76 | + 33 | + 27 |

30

F
| 61 | 67 | 36 | 21 | 35 | 10 |
| + 69 | + 21 | + 27 | + 30 | + 40 | + 85 |

G
| 96 | 16 | 93 | 37 | 59 | 20 |
| + 16 | + 16 | + 87 | + 14 | + 25 | + 30 |

H
| 46 | 75 | 21 | 28 | 86 | 24 |
| + 28 | + 88 | + 42 | + 70 | + 91 | + 38 |

I
| 73 | 65 | 29 | 48 | 17 | 39 |
| + 83 | + 99 | + 27 | + 87 | + 14 | + 43 |

J
| 29 | 79 | 73 | 33 | 37 | 11 |
| + 83 | + 15 | + 44 | + 25 | + 89 | + 65 |

60

A
```
  13      70      93      19      76      85
+ 54    + 94    + 92    + 42    + 41    + 81
```

B
```
  78      65      74      19      50      32
+ 30    + 49    + 33    + 78    + 94    + 37
```

C
```
  88      99      73      27      13      33
+ 46    + 97    + 46    + 51    + 51    + 29
```

D
```
  31      73      34      81      52      34
+ 82    + 33    + 39    + 72    + 56    + 20
```

E
```
  75      10      70      14      88      45
+ 82    + 59    + 45    + 64    + 32    + 85
```

F
```
  10      35      21      87      20      67
+ 51    + 22    + 79    + 14    + 55    + 28
```

G
```
  54      80      25      23      82      70
+ 21    + 34    + 19    + 59    + 38    + 95
```

H
```
  44      99      91      38      10      28
+ 41    + 24    + 29    + 44    + 97    + 87
```

I
```
  35      25      63      93      64      45
+ 13    + 83    + 33    + 24    + 37    + 32
```

J
```
  21      34      46      85      56      53
+ 92    + 47    + 80    + 51    + 75    + 32
```

A
| 54 | 59 | 74 | 68 | 86 | 35 |
| + 66 | + 42 | + 83 | + 35 | + 24 | + 14 |

B
| 76 | 92 | 48 | 66 | 98 | 78 |
| + 62 | + 46 | + 20 | + 59 | + 63 | + 82 |

C
| 36 | 62 | 89 | 46 | 12 | 37 |
| + 58 | + 51 | + 97 | + 17 | + 28 | + 82 |

D
| 31 | 37 | 25 | 24 | 98 | 14 |
| + 51 | + 12 | + 73 | + 16 | + 12 | + 21 |

E
| 25 | 80 | 14 | 53 | 15 | 87 |
| + 60 | + 49 | + 22 | + 86 | + 94 | + 28 |

F
| 42 | 73 | 54 | 88 | 88 | 60 |
| + 70 | + 97 | + 28 | + 72 | + 17 | + 38 |

G
| 67 | 54 | 23 | 51 | 23 | 70 |
| + 77 | + 15 | + 86 | + 52 | + 52 | + 60 |

H
| 59 | 76 | 21 | 10 | 72 | 90 |
| + 74 | + 33 | + 38 | + 57 | + 45 | + 41 |

I
| 69 | 62 | 12 | 62 | 25 | 73 |
| + 31 | + 88 | + 12 | + 44 | + 99 | + 14 |

J
| 11 | 38 | 77 | 62 | 56 | 54 |
| + 58 | + 91 | + 40 | + 38 | + 22 | + 86 |

A
| 90 | 95 | 60 | 35 | 39 | 56 |
| + 59 | + 50 | + 51 | + 76 | + 58 | + 98 |

B
| 77 | 88 | 76 | 60 | 65 | 66 |
| + 35 | + 25 | + 32 | + 79 | + 99 | + 39 |

C
| 11 | 23 | 17 | 24 | 17 | 12 |
| + 33 | + 57 | + 50 | + 74 | + 46 | + 46 |

D
| 74 | 73 | 46 | 39 | 47 | 73 |
| + 12 | + 33 | + 37 | + 93 | + 70 | + 89 |

E
| 46 | 37 | 73 | 79 | 16 | 77 |
| + 64 | + 85 | + 23 | + 98 | + 47 | + 55 |

F
| 75 | 35 | 79 | 66 | 44 | 94 |
| + 12 | + 11 | + 18 | + 26 | + 34 | + 11 |

G
| 87 | 26 | 91 | 87 | 48 | 52 |
| + 27 | + 76 | + 99 | + 43 | + 32 | + 94 |

H
| 26 | 57 | 87 | 54 | 19 | 41 |
| + 51 | + 61 | + 78 | + 40 | + 29 | + 77 |

I
| 92 | 60 | 15 | 34 | 25 | 71 |
| + 46 | + 77 | + 45 | + 92 | + 29 | + 48 |

J
| 18 | 50 | 31 | 38 | 28 | 26 |
| + 68 | + 20 | + 52 | + 55 | + 84 | + 79 |

30

60

A
$$75 + 60$$ $$55 + 91$$ $$27 + 17$$ $$94 + 86$$ $$29 + 12$$ $$67 + 81$$

B
$$61 + 67$$ $$82 + 56$$ $$99 + 22$$ $$33 + 31$$ $$62 + 68$$ $$27 + 59$$

C
$$14 + 91$$ $$11 + 74$$ $$49 + 93$$ $$65 + 55$$ $$78 + 81$$ $$35 + 15$$

D
$$22 + 44$$ $$43 + 26$$ $$67 + 76$$ $$49 + 86$$ $$79 + 38$$ $$33 + 39$$

E
$$56 + 79$$ $$89 + 27$$ $$28 + 17$$ $$89 + 49$$ $$21 + 67$$ $$88 + 65$$

F
$$31 + 37$$ $$15 + 41$$ $$36 + 34$$ $$76 + 91$$ $$21 + 90$$ $$79 + 89$$

G
$$59 + 77$$ $$68 + 14$$ $$18 + 34$$ $$59 + 78$$ $$72 + 50$$ $$22 + 81$$

H
$$50 + 21$$ $$80 + 72$$ $$26 + 54$$ $$21 + 51$$ $$63 + 63$$ $$16 + 75$$

I
$$50 + 74$$ $$76 + 26$$ $$53 + 11$$ $$45 + 97$$ $$16 + 65$$ $$87 + 42$$

J
$$43 + 78$$ $$44 + 61$$ $$10 + 57$$ $$38 + 40$$ $$28 + 74$$ $$94 + 83$$

A	56 + 33	66 + 85	72 + 39	80 + 65	23 + 15	82 + 13
B	66 + 79	85 + 19	97 + 43	63 + 89	25 + 87	88 + 26
C	75 + 34	77 + 47	68 + 78	27 + 64	93 + 71	66 + 71
D	41 + 60	62 + 23	97 + 23	52 + 11	53 + 78	35 + 49
E	60 + 64	22 + 96	19 + 83	29 + 53	43 + 47	66 + 29
F	58 + 59	23 + 18	16 + 84	68 + 38	66 + 21	50 + 31
G	19 + 98	73 + 97	33 + 51	13 + 68	40 + 62	87 + 36
H	18 + 62	26 + 91	72 + 68	44 + 56	50 + 16	65 + 70
I	33 + 12	61 + 57	43 + 89	71 + 12	73 + 22	35 + 93
J	43 + 29	73 + 26	67 + 45	37 + 43	87 + 66	77 + 74

30

60

A

81	37	39	93	24	66
+ 49	+ 65	+ 18	+ 24	+ 40	+ 87

B

14	18	86	76	15	64
+ 81	+ 96	+ 43	+ 45	+ 25	+ 18

C

19	38	35	90	66	62
+ 62	+ 43	+ 61	+ 39	+ 99	+ 95

D

35	14	95	66	85	14
+ 57	+ 34	+ 96	+ 14	+ 81	+ 25

E

92	44	14	40	11	32
+ 64	+ 65	+ 56	+ 92	+ 99	+ 22

F

50	37	17	18	36	33
+ 89	+ 44	+ 37	+ 12	+ 84	+ 20

G

75	28	90	97	49	77
+ 58	+ 15	+ 43	+ 72	+ 34	+ 72

H

14	95	29	33	34	64
+ 57	+ 83	+ 60	+ 89	+ 62	+ 24

I

13	15	45	30	47	80
+ 89	+ 89	+ 97	+ 16	+ 38	+ 49

J

72	10	43	20	26	99
+ 95	+ 64	+ 71	+ 41	+ 14	+ 40

A
| 63 | 69 | 79 | 64 | 49 | 37 |
| + 59 | + 25 | + 33 | + 29 | + 90 | + 62 |

B
| 17 | 80 | 19 | 30 | 46 | 80 |
| + 68 | + 70 | + 78 | + 91 | + 35 | + 18 |

C
| 80 | 33 | 65 | 33 | 13 | 38 |
| + 17 | + 75 | + 78 | + 87 | + 36 | + 57 |

D
| 94 | 73 | 95 | 95 | 90 | 76 |
| + 95 | + 49 | + 70 | + 45 | + 65 | + 82 |

E
| 88 | 17 | 47 | 72 | 58 | 21 |
| + 43 | + 72 | + 25 | + 77 | + 99 | + 18 |

F
| 53 | 76 | 70 | 84 | 16 | 98 |
| + 38 | + 75 | + 30 | + 55 | + 26 | + 98 |

G
| 96 | 91 | 96 | 61 | 56 | 58 |
| + 55 | + 40 | + 64 | + 13 | + 19 | + 12 |

H
| 16 | 22 | 80 | 76 | 66 | 51 |
| + 96 | + 98 | + 63 | + 55 | + 20 | + 28 |

I
| 27 | 23 | 22 | 97 | 97 | 56 |
| + 75 | + 77 | + 13 | + 78 | + 97 | + 17 |

J
| 50 | 64 | 82 | 55 | 49 | 49 |
| + 21 | + 59 | + 96 | + 69 | + 49 | + 97 |

A

44	59	25	28	28	91
+ 41	+ 19	+ 75	+ 46	+ 80	+ 43

B

15	80	93	39	65	18
+ 74	+ 70	+ 89	+ 77	+ 30	+ 76

C

49	77	88	56	84	95
+ 35	+ 75	+ 69	+ 65	+ 24	+ 95

D

73	48	16	54	21	27
+ 47	+ 47	+ 31	+ 38	+ 44	+ 24

E

39	82	85	28	57	71
+ 19	+ 57	+ 57	+ 45	+ 61	+ 31

F

80	35	38	15	27	35
+ 42	+ 50	+ 33	+ 15	+ 16	+ 27

G

82	16	57	87	35	41
+ 48	+ 12	+ 65	+ 23	+ 97	+ 54

H

12	76	11	76	31	83
+ 40	+ 34	+ 83	+ 46	+ 49	+ 81

I

62	87	72	75	57	45
+ 97	+ 55	+ 91	+ 25	+ 38	+ 97

J

97	28	43	58	26	28
+ 58	+ 44	+ 68	+ 87	+ 38	+ 41

A
25	45	65	71	66	34
+ 65	+ 42	+ 11	+ 35	+ 91	+ 17

B
53	10	24	22	73	66
+ 53	+ 44	+ 87	+ 68	+ 47	+ 18

C
23	27	98	96	72	21
+ 95	+ 34	+ 33	+ 78	+ 35	+ 68

D
70	51	24	15	64	82
+ 42	+ 20	+ 64	+ 21	+ 48	+ 18

E
53	78	37	15	48	57
+ 60	+ 84	+ 46	+ 86	+ 36	+ 42

F
42	33	93	49	23	23
+ 75	+ 70	+ 49	+ 53	+ 75	+ 96

G
39	38	36	30	21	26
+ 89	+ 36	+ 32	+ 75	+ 53	+ 33

H
89	81	66	64	32	31
+ 98	+ 56	+ 47	+ 15	+ 95	+ 17

I
85	67	60	78	75	17
+ 87	+ 65	+ 40	+ 69	+ 27	+ 81

J
23	95	34	41	81	51
+ 47	+ 49	+ 76	+ 67	+ 66	+ 71

A	52 + 58	53 + 46	28 + 10	82 + 97	96 + 68	27 + 95
B	85 + 42	54 + 27	94 + 25	11 + 63	30 + 95	56 + 58
C	32 + 83	40 + 95	66 + 55	94 + 39	47 + 49	25 + 81
D	35 + 33	30 + 41	47 + 59	88 + 23	75 + 62	77 + 41
E	90 + 21	11 + 93	40 + 17	68 + 34	94 + 71	94 + 99
F	93 + 93	59 + 37	17 + 27	18 + 90	30 + 16	56 + 86
G	75 + 35	25 + 32	97 + 68	41 + 29	11 + 68	31 + 33
H	43 + 77	97 + 63	38 + 83	20 + 23	24 + 26	65 + 69
I	46 + 12	71 + 57	54 + 37	93 + 47	60 + 27	24 + 52
J	79 + 29	15 + 99	34 + 69	47 + 30	17 + 70	81 + 83

A

90	44	72	92	48	11
+ 98	+ 81	+ 51	+ 33	+ 54	+ 31

B

82	94	47	11	37	53
+ 40	+ 90	+ 82	+ 32	+ 44	+ 16

C

59	65	47	72	40	82
+ 67	+ 13	+ 93	+ 68	+ 65	+ 50

D

65	39	80	75	16	57
+ 48	+ 18	+ 46	+ 23	+ 79	+ 98

E

75	36	51	93	72	77
+ 64	+ 39	+ 65	+ 33	+ 51	+ 98

F

74	90	58	26	78	77
+ 53	+ 93	+ 49	+ 80	+ 18	+ 75

G

48	50	35	82	66	57
+ 43	+ 18	+ 60	+ 31	+ 91	+ 37

H

72	36	61	63	90	70
+ 34	+ 22	+ 94	+ 82	+ 58	+ 27

I

44	67	61	94	91	59
+ 90	+ 41	+ 72	+ 98	+ 95	+ 88

J

31	95	49	86	81	10
+ 33	+ 22	+ 91	+ 44	+ 73	+ 62

A

31	55	27	15	45	50
+ 76	+ 71	+ 78	+ 40	+ 30	+ 40

B

39	63	34	72	24	32
+ 93	+ 67	+ 20	+ 41	+ 76	+ 92

C

11	86	89	62	17	27
+ 66	+ 39	+ 43	+ 88	+ 39	+ 38

D

26	21	15	32	23	92
+ 69	+ 42	+ 38	+ 77	+ 91	+ 87

E

17	97	77	70	35	45
+ 20	+ 17	+ 93	+ 59	+ 38	+ 40

F

66	34	50	18	34	93
+ 55	+ 84	+ 84	+ 46	+ 11	+ 51

G

31	67	53	26	41	47
+ 31	+ 31	+ 91	+ 61	+ 63	+ 77

H

19	31	13	42	95	64
+ 89	+ 66	+ 48	+ 10	+ 16	+ 91

I

66	39	81	78	51	55
+ 48	+ 31	+ 65	+ 92	+ 92	+ 34

J

72	98	42	99	98	44
+ 75	+ 71	+ 60	+ 65	+ 30	+ 30

Math Blank Worksheet

A

83	72	82	49	73	68
− 52	− 53	− 24	− 15	− 44	− 28

B

73	25	63	81	88	98
− 55	− 11	− 16	− 37	− 31	− 90

C

99	41	91	81	48	42
− 53	− 30	− 16	− 63	− 16	− 22

D

83	95	82	93	69	64
− 41	− 36	− 63	− 71	− 45	− 13

E

16	76	84	99	81	63
− 15	− 41	− 39	− 65	− 69	− 42

F

66	87	35	71	69	83
− 64	− 36	− 26	− 59	− 50	− 17

G

87	95	96	66	81	50
− 64	− 30	− 90	− 57	− 54	− 30

H

27	82	79	62	47	18
− 23	− 25	− 77	− 37	− 45	− 13

I

60	54	42	94	35	25
− 38	− 29	− 30	− 59	− 19	− 14

J

90	55	70	60	62	45
− 41	− 23	− 47	− 38	− 14	− 19

A

68	28	74	54	80	96
− 27	− 28	− 19	− 23	− 67	− 49

B

85	73	57	53	52	99
− 79	− 54	− 52	− 22	− 48	− 34

C

98	97	65	98	72	47
− 72	− 47	− 33	− 11	− 71	− 24

D

18	70	81	65	27	30
− 16	− 32	− 73	− 23	− 15	− 14

E

74	46	47	78	74	50
− 11	− 23	− 14	− 19	− 29	− 17

30

F

90	33	91	79	75	32
− 41	− 11	− 53	− 24	− 73	− 27

G

20	40	69	78	87	72
− 17	− 22	− 46	− 29	− 46	− 69

H

56	78	51	88	57	15
− 43	− 74	− 20	− 38	− 32	− 15

I

43	55	58	44	94	42
− 42	− 27	− 32	− 42	− 30	− 11

J

54	82	96	71	35	43
− 17	− 25	− 49	− 18	− 12	− 35

60

A
| 87 | 53 | 98 | 87 | 72 | 63 |
| − 37 | − 13 | − 20 | − 60 | − 50 | − 11 |

B
| 47 | 41 | 87 | 56 | 67 | 65 |
| − 15 | − 29 | − 67 | − 36 | − 42 | − 54 |

C
| 97 | 58 | 70 | 93 | 78 | 66 |
| − 47 | − 43 | − 32 | − 45 | − 20 | − 53 |

D
| 42 | 59 | 72 | 77 | 99 | 72 |
| − 21 | − 45 | − 39 | − 77 | − 97 | − 45 |

E
| 73 | 99 | 47 | 98 | 44 | 69 |
| − 34 | − 18 | − 42 | − 59 | − 31 | − 53 |

F
| 45 | 77 | 87 | 87 | 18 | 88 |
| − 12 | − 21 | − 42 | − 87 | − 13 | − 36 |

G
| 82 | 79 | 71 | 93 | 20 | 91 |
| − 37 | − 18 | − 49 | − 31 | − 11 | − 15 |

H
| 51 | 85 | 60 | 77 | 93 | 68 |
| − 32 | − 51 | − 13 | − 29 | − 24 | − 28 |

I
| 97 | 95 | 98 | 82 | 97 | 84 |
| − 37 | − 40 | − 69 | − 12 | − 36 | − 41 |

J
| 91 | 72 | 84 | 89 | 90 | 32 |
| − 87 | − 58 | − 70 | − 80 | − 21 | − 27 |

A	43 − 23	41 − 25	87 − 17	83 − 63	89 − 19	68 − 68
B	65 − 47	34 − 14	97 − 94	92 − 29	80 − 49	30 − 22
C	84 − 56	96 − 42	97 − 24	52 − 43	93 − 12	49 − 36
D	60 − 43	83 − 72	77 − 75	84 − 56	70 − 61	91 − 46
E	63 − 32	59 − 54	70 − 50	82 − 36	25 − 17	98 − 11
F	76 − 71	51 − 21	80 − 36	90 − 40	46 − 45	67 − 65
G	81 − 11	52 − 26	37 − 23	96 − 29	71 − 16	25 − 17
H	79 − 16	76 − 75	96 − 31	89 − 32	69 − 33	62 − 14
I	81 − 13	53 − 14	87 − 42	75 − 27	83 − 47	98 − 15
J	75 − 12	74 − 69	23 − 17	87 − 73	53 − 50	88 − 81

30

60

A
55	43	64	77	81	64
− 45	− 11	− 42	− 31	− 16	− 24

B
30	58	81	97	97	87
− 27	− 31	− 43	− 21	− 54	− 33

C
81	84	69	16	46	82
− 46	− 80	− 10	− 12	− 28	− 48

D
68	91	63	92	96	47
− 17	− 53	− 41	− 45	− 73	− 40

E
97	64	80	70	82	98
− 79	− 60	− 64	− 25	− 30	− 87

F
60	86	63	81	57	38
− 31	− 44	− 63	− 22	− 39	− 34

G
32	54	47	90	47	53
− 10	− 54	− 25	− 49	− 46	− 14

H
82	89	27	35	19	51
− 28	− 43	− 20	− 30	− 17	− 17

I
46	67	74	74	49	64
− 13	− 39	− 46	− 57	− 10	− 36

J
97	99	99	43	71	88
− 55	− 94	− 34	− 19	− 17	− 44

30

60

A
| 79 | 79 | 90 | 83 | 42 | 43 |
| − 76 | − 73 | − 86 | − 71 | − 20 | − 26 |

B
| 44 | 13 | 62 | 59 | 83 | 19 |
| − 35 | − 10 | − 54 | − 23 | − 76 | − 11 |

C
| 76 | 25 | 98 | 71 | 95 | 70 |
| − 18 | − 23 | − 51 | − 10 | − 12 | − 60 |

D
| 58 | 88 | 60 | 91 | 22 | 43 |
| − 45 | − 81 | − 38 | − 21 | − 15 | − 34 |

E
| 19 | 94 | 83 | 56 | 85 | 95 |
| − 18 | − 85 | − 72 | − 15 | − 76 | − 53 |

30

F
| 64 | 46 | 93 | 37 | 29 | 75 |
| − 28 | − 39 | − 31 | − 28 | − 21 | − 11 |

G
| 38 | 25 | 36 | 54 | 44 | 90 |
| − 35 | − 20 | − 33 | − 22 | − 17 | − 69 |

H
| 34 | 61 | 88 | 65 | 80 | 66 |
| − 12 | − 35 | − 87 | − 53 | − 42 | − 33 |

I
| 47 | 84 | 28 | 72 | 81 | 98 |
| − 13 | − 45 | − 25 | − 50 | − 23 | − 84 |

J
| 99 | 86 | 71 | 89 | 68 | 44 |
| − 71 | − 33 | − 52 | − 12 | − 41 | − 17 |

60

A
99	67	99	82	71	47
− 29	− 55	− 26	− 72	− 64	− 38

B
93	63	75	28	74	86
− 61	− 29	− 41	− 10	− 32	− 35

C
53	92	69	89	86	43
− 52	− 31	− 53	− 83	− 79	− 13

D
99	44	96	77	92	31
− 60	− 26	− 70	− 38	− 50	− 19

E
99	99	36	74	73	51
− 43	− 18	− 32	− 46	− 66	− 33

F
91	89	82	77	84	74
− 75	− 16	− 65	− 10	− 63	− 67

G
29	48	76	94	81	76
− 17	− 17	− 46	− 37	− 37	− 11

H
91	91	68	34	87	78
− 52	− 50	− 39	− 18	− 22	− 39

I
95	64	92	92	73	27
− 89	− 21	− 55	− 91	− 37	− 26

J
87	78	89	91	70	46
− 63	− 72	− 27	− 36	− 70	− 30

A	94 − 87	99 − 43	98 − 18	74 − 31	69 − 33	68 − 11
B	79 − 19	73 − 40	98 − 97	35 − 26	63 − 35	55 − 53
C	89 − 70	90 − 25	75 − 44	64 − 34	52 − 27	87 − 70
D	71 − 43	86 − 46	73 − 61	68 − 53	86 − 40	73 − 10
E	82 − 10	90 − 80	98 − 74	88 − 73	57 − 13	92 − 84
F	91 − 71	60 − 56	78 − 43	88 − 11	82 − 49	23 − 22
G	99 − 67	98 − 66	81 − 71	64 − 47	75 − 56	40 − 19
H	60 − 53	87 − 38	80 − 17	67 − 26	78 − 39	77 − 73
I	72 − 14	35 − 11	93 − 80	99 − 77	70 − 69	91 − 77
J	61 − 58	35 − 34	34 − 11	69 − 44	84 − 20	43 − 12

30

60

A

| 17
− 11 | 66
− 50 | 51
− 31 | 58
− 24 | 62
− 36 | 94
− 37 |

B

| 40
− 28 | 69
− 28 | 68
− 34 | 87
− 36 | 54
− 44 | 78
− 48 |

C

| 80
− 28 | 33
− 21 | 82
− 23 | 99
− 27 | 41
− 10 | 59
− 32 |

D

| 65
− 45 | 84
− 11 | 19
− 18 | 87
− 32 | 57
− 10 | 67
− 49 |

E

| 89
− 78 | 67
− 63 | 47
− 40 | 83
− 82 | 57
− 53 | 21
− 10 |

F

| 32
− 30 | 49
− 11 | 66
− 62 | 96
− 93 | 83
− 57 | 97
− 80 |

G

| 44
− 12 | 97
− 85 | 52
− 29 | 57
− 47 | 80
− 24 | 62
− 51 |

H

| 16
− 14 | 45
− 25 | 76
− 34 | 88
− 51 | 31
− 24 | 64
− 42 |

I

| 93
− 17 | 38
− 34 | 57
− 16 | 48
− 37 | 98
− 53 | 39
− 27 |

J

| 97
− 17 | 66
− 16 | 66
− 61 | 42
− 25 | 50
− 29 | 87
− 43 |

A

78	75	95	66	15	53
− 35	− 65	− 33	− 59	− 11	− 15

B

47	91	42	23	60	40
− 31	− 40	− 30	− 20	− 51	− 10

C

90	90	64	47	46	26
− 50	− 64	− 21	− 26	− 12	− 25

D

65	28	20	17	79	64
− 62	− 14	− 17	− 15	− 67	− 62

E

63	90	53	70	38	73
− 31	− 27	− 47	− 23	− 15	− 59

F

82	79	74	61	72	78
− 19	− 43	− 31	− 51	− 44	− 65

G

97	71	65	71	50	87
− 15	− 29	− 11	− 29	− 24	− 42

H

34	44	51	62	66	79
− 15	− 24	− 40	− 54	− 26	− 36

I

52	94	53	90	28	51
− 50	− 67	− 16	− 25	− 11	− 25

J

46	55	49	65	43	95
− 45	− 27	− 45	− 33	− 25	− 56

A

26	77	79	59	84	58
− 25	− 69	− 18	− 18	− 15	− 16

B

84	66	45	53	90	19
− 67	− 42	− 37	− 27	− 49	− 11

C

95	45	98	57	83	84
− 68	− 27	− 37	− 33	− 19	− 11

D

95	82	99	81	65	65
− 65	− 77	− 55	− 25	− 55	− 34

E

61	86	74	50	44	89
− 39	− 13	− 38	− 26	− 14	− 73

F

95	89	93	28	89	62
− 30	− 43	− 48	− 14	− 23	− 54

G

55	78	81	98	67	74
− 20	− 75	− 63	− 91	− 32	− 60

H

85	87	79	72	44	93
− 30	− 42	− 53	− 49	− 41	− 24

I

39	84	75	61	20	92
− 29	− 83	− 56	− 34	− 13	− 75

J

80	32	59	56	68	47
− 10	− 13	− 13	− 27	− 12	− 29

A

55	98	38	91	64	44
− 53	− 21	− 13	− 66	− 55	− 22

B

34	71	83	97	89	64
− 12	− 12	− 40	− 27	− 83	− 53

C

65	25	96	91	93	67
− 29	− 23	− 67	− 23	− 30	− 31

D

90	88	87	96	95	99
− 76	− 51	− 15	− 59	− 61	− 27

E

59	91	89	43	93	77
− 14	− 69	− 16	− 22	− 80	− 55

30

F

41	54	76	28	61	89
− 35	− 11	− 28	− 14	− 11	− 86

G

77	94	52	98	69	76
− 46	− 71	− 43	− 95	− 50	− 73

H

82	54	89	66	87	25
− 27	− 21	− 32	− 10	− 59	− 22

I

56	86	76	45	43	60
− 16	− 34	− 50	− 23	− 20	− 26

J

26	71	24	29	99	80
− 16	− 34	− 18	− 20	− 84	− 21

60

A
| 67 | 56 | 74 | 40 | 55 | 87 |
| - 65 | - 45 | - 30 | - 29 | - 29 | - 32 |

B
| 89 | 74 | 73 | 55 | 20 | 50 |
| - 21 | - 54 | - 38 | - 32 | - 15 | - 17 |

C
| 96 | 53 | 97 | 90 | 36 | 44 |
| - 23 | - 39 | - 88 | - 90 | - 13 | - 11 |

D
| 53 | 83 | 85 | 92 | 97 | 16 |
| - 37 | - 24 | - 17 | - 53 | - 16 | - 10 |

E
| 95 | 81 | 89 | 79 | 73 | 77 |
| - 78 | - 76 | - 28 | - 38 | - 56 | - 25 |

F
| 92 | 68 | 77 | 77 | 33 | 57 |
| - 24 | - 41 | - 72 | - 57 | - 13 | - 55 |

G
| 33 | 96 | 99 | 97 | 47 | 79 |
| - 31 | - 71 | - 38 | - 75 | - 28 | - 61 |

H
| 20 | 94 | 64 | 73 | 32 | 72 |
| - 18 | - 53 | - 56 | - 62 | - 26 | - 26 |

I
| 89 | 57 | 44 | 99 | 49 | 90 |
| - 50 | - 56 | - 36 | - 72 | - 12 | - 31 |

J
| 96 | 80 | 37 | 62 | 72 | 95 |
| - 40 | - 54 | - 32 | - 46 | - 27 | - 56 |

A

62	52	33	91	84	34
− 47	− 47	− 31	− 87	− 72	− 12

B

94	90	98	44	22	79
− 36	− 58	− 48	− 34	− 17	− 58

C

99	37	36	62	51	78
− 97	− 18	− 12	− 48	− 42	− 49

D

48	83	98	29	62	87
− 37	− 12	− 84	− 25	− 54	− 38

E

73	59	57	99	85	78
− 43	− 32	− 39	− 29	− 78	− 47

30

F

40	73	59	58	96	64
− 37	− 36	− 39	− 21	− 43	− 12

G

63	15	50	33	95	77
− 58	− 11	− 43	− 15	− 33	− 44

H

74	43	38	91	87	61
− 43	− 29	− 24	− 50	− 13	− 29

I

96	50	56	65	91	26
− 87	− 44	− 52	− 43	− 17	− 16

J

77	40	38	63	49	98
− 37	− 21	− 21	− 48	− 17	− 98

60

A
```
  87       97       91       80       91       64
- 46     - 44     - 67     - 69     - 58     - 22
```

B
```
  89       42       36       75       57       55
- 25     - 26     - 31     - 15     - 26     - 54
```

C
```
  85       41       82       81       66       87
- 83     - 39     - 64     - 70     - 48     - 66
```

D
```
  88       92       64       70       13       37
- 28     - 28     - 21     - 61     - 10     - 12
```

E
```
  70       52       95       23       47       37
- 20     - 29     - 39     - 17     - 39     - 31
```

F
```
  78       83       70       57       63       87
- 63     - 70     - 45     - 45     - 24     - 74
```

G
```
  77       82       60       85       38       50
- 40     - 58     - 31     - 29     - 25     - 14
```

H
```
  93       27       15       76       96       32
- 62     - 26     - 11     - 61     - 48     - 24
```

I
```
  62       51       95       42       97       79
- 31     - 34     - 52     - 10     - 37     - 43
```

J
```
  93       28       91       91       33       92
- 71     - 16     - 48     - 60     - 32     - 13
```

A
| 84 | 42 | 49 | 86 | 76 | 78 |
| − 48 | − 32 | − 40 | − 72 | − 20 | − 15 |

B
| 95 | 97 | 48 | 44 | 75 | 99 |
| − 66 | − 39 | − 43 | − 35 | − 62 | − 63 |

C
| 99 | 44 | 34 | 89 | 77 | 53 |
| − 83 | − 23 | − 11 | − 13 | − 35 | − 49 |

D
| 95 | 93 | 95 | 81 | 58 | 51 |
| − 19 | − 72 | − 33 | − 79 | − 53 | − 12 |

E
| 99 | 59 | 80 | 67 | 53 | 93 |
| − 88 | − 23 | − 59 | − 13 | − 51 | − 82 |

F
| 84 | 64 | 45 | 35 | 89 | 71 |
| − 15 | − 19 | − 42 | − 10 | − 44 | − 61 |

G
| 55 | 58 | 63 | 94 | 23 | 21 |
| − 13 | − 40 | − 50 | − 55 | − 16 | − 20 |

H
| 55 | 97 | 89 | 29 | 60 | 95 |
| − 12 | − 96 | − 84 | − 18 | − 55 | − 13 |

I
| 56 | 95 | 69 | 29 | 99 | 87 |
| − 53 | − 49 | − 65 | − 16 | − 56 | − 15 |

J
| 73 | 36 | 95 | 85 | 96 | 70 |
| − 11 | − 18 | − 62 | − 63 | − 27 | − 30 |

A	25 − 10	68 − 38	62 − 33	93 − 64	93 − 70	61 − 45
B	66 − 63	87 − 12	81 − 13	97 − 17	97 − 67	72 − 25
C	73 − 14	64 − 59	44 − 24	61 − 23	31 − 17	64 − 18
D	97 − 18	58 − 37	92 − 15	96 − 18	84 − 29	95 − 69
E	62 − 20	90 − 28	87 − 83	59 − 30	70 − 33	49 − 33
F	61 − 35	90 − 69	66 − 12	93 − 70	72 − 40	27 − 21
G	79 − 39	86 − 11	94 − 12	78 − 44	96 − 83	94 − 64
H	55 − 23	66 − 61	90 − 57	30 − 18	35 − 18	89 − 76
I	89 − 40	93 − 10	60 − 53	72 − 69	96 − 48	94 − 54
J	30 − 27	39 − 36	92 − 25	68 − 34	86 − 37	26 − 17

30

60

A
| 17 | 88 | 96 | 56 | 32 | 81 |
| − 15 | − 35 | − 28 | − 27 | − 26 | − 10 |

B
| 55 | 57 | 53 | 62 | 53 | 53 |
| − 34 | − 32 | − 39 | − 59 | − 16 | − 34 |

C
| 95 | 78 | 46 | 69 | 69 | 80 |
| − 69 | − 46 | − 12 | − 20 | − 19 | − 17 |

D
| 82 | 82 | 88 | 50 | 98 | 26 |
| − 58 | − 65 | − 48 | − 16 | − 42 | − 23 |

E
| 96 | 95 | 82 | 82 | 78 | 70 |
| − 48 | − 42 | − 13 | − 13 | − 66 | − 26 |

F
| 33 | 68 | 95 | 96 | 30 | 84 |
| − 30 | − 11 | − 90 | − 78 | − 27 | − 60 |

G
| 60 | 79 | 83 | 61 | 43 | 86 |
| − 30 | − 28 | − 79 | − 37 | − 24 | − 26 |

H
| 93 | 21 | 83 | 72 | 81 | 95 |
| − 67 | − 11 | − 25 | − 59 | − 36 | − 32 |

I
| 91 | 16 | 94 | 81 | 68 | 68 |
| − 56 | − 10 | − 88 | − 71 | − 21 | − 43 |

J
| 70 | 90 | 92 | 90 | 35 | 79 |
| − 49 | − 46 | − 44 | − 41 | − 11 | − 17 |

A
| 45 | 50 | 87 | 80 | 52 | 44 |
| − 30 | − 25 | − 70 | − 66 | − 24 | − 17 |

B
| 65 | 64 | 57 | 82 | 85 | 48 |
| − 53 | − 14 | − 30 | − 11 | − 24 | − 20 |

C
| 36 | 65 | 58 | 51 | 90 | 92 |
| − 12 | − 19 | − 51 | − 24 | − 42 | − 16 |

D
| 59 | 97 | 63 | 83 | 50 | 29 |
| − 12 | − 73 | − 60 | − 39 | − 14 | − 22 |

E
| 58 | 51 | 75 | 74 | 91 | 98 |
| − 14 | − 22 | − 26 | − 57 | − 51 | − 41 |

F
| 56 | 97 | 60 | 97 | 23 | 77 |
| − 17 | − 43 | − 19 | − 72 | − 16 | − 39 |

G
| 81 | 93 | 97 | 88 | 36 | 76 |
| − 21 | − 28 | − 17 | − 78 | − 15 | − 67 |

H
| 95 | 70 | 48 | 70 | 58 | 96 |
| − 36 | − 18 | − 22 | − 61 | − 28 | − 92 |

I
| 34 | 59 | 42 | 77 | 45 | 78 |
| − 27 | − 14 | − 26 | − 41 | − 23 | − 36 |

J
| 72 | 55 | 79 | 66 | 94 | 53 |
| − 27 | − 39 | − 29 | − 16 | − 76 | − 14 |

A

| 90
− 58 | 75
− 55 | 43
− 42 | 30
− 15 | 76
− 28 | 99
− 24 |

B

| 87
− 84 | 86
− 30 | 62
− 55 | 78
− 58 | 48
− 23 | 37
− 30 |

C

| 58
− 44 | 55
− 16 | 64
− 55 | 76
− 57 | 74
− 10 | 82
− 22 |

D

| 68
− 52 | 99
− 37 | 42
− 12 | 42
− 13 | 58
− 57 | 84
− 79 |

E

| 95
− 84 | 51
− 21 | 88
− 82 | 92
− 90 | 67
− 64 | 40
− 14 |

F

| 97
− 39 | 33
− 21 | 31
− 20 | 95
− 42 | 84
− 31 | 78
− 68 |

G

| 85
− 10 | 64
− 42 | 63
− 53 | 37
− 19 | 78
− 48 | 87
− 23 |

H

| 67
− 55 | 99
− 89 | 46
− 25 | 90
− 64 | 92
− 25 | 94
− 16 |

I

| 90
− 48 | 88
− 67 | 71
− 19 | 49
− 12 | 58
− 47 | 59
− 42 |

J

| 58
− 50 | 94
− 85 | 54
− 24 | 82
− 82 | 99
− 43 | 43
− 35 |

30

60

A
| 85 | 93 | 94 | 98 | 30 | 83 |
| − 46 | − 75 | − 50 | − 16 | − 17 | − 75 |

B
| 84 | 64 | 43 | 58 | 84 | 47 |
| − 15 | − 38 | − 20 | − 26 | − 60 | − 40 |

C
| 94 | 73 | 78 | 28 | 87 | 91 |
| − 63 | − 60 | − 55 | − 24 | − 29 | − 46 |

D
| 40 | 79 | 98 | 57 | 12 | 22 |
| − 32 | − 68 | − 51 | − 40 | − 11 | − 14 |

E
| 67 | 87 | 65 | 59 | 97 | 83 |
| − 50 | − 78 | − 61 | − 23 | − 40 | − 30 |

F
| 73 | 97 | 85 | 19 | 51 | 92 |
| − 27 | − 95 | − 59 | − 18 | − 11 | − 11 |

G
| 60 | 31 | 52 | 60 | 89 | 89 |
| − 11 | − 13 | − 38 | − 29 | − 49 | − 85 |

H
| 61 | 90 | 91 | 85 | 66 | 70 |
| − 43 | − 11 | − 70 | − 78 | − 38 | − 33 |

I
| 68 | 64 | 60 | 76 | 52 | 72 |
| − 12 | − 16 | − 52 | − 56 | − 33 | − 40 |

J
| 95 | 73 | 73 | 34 | 34 | 55 |
| − 78 | − 28 | − 30 | − 27 | − 26 | − 41 |

A
| 37 | 84 | 59 | 87 | 67 | 73 |
| − 10 | − 69 | − 20 | − 47 | − 62 | − 39 |

B
| 81 | 94 | 74 | 93 | 70 | 38 |
| − 80 | − 39 | − 26 | − 48 | − 22 | − 10 |

C
| 46 | 82 | 84 | 84 | 92 | 50 |
| − 22 | − 47 | − 38 | − 35 | − 32 | − 42 |

D
| 96 | 81 | 87 | 30 | 85 | 61 |
| − 19 | − 26 | − 11 | − 20 | − 39 | − 19 |

E
| 72 | 83 | 33 | 65 | 64 | 41 |
| − 51 | − 80 | − 14 | − 64 | − 61 | − 38 |

F
| 85 | 51 | 84 | 62 | 70 | 60 |
| − 85 | − 30 | − 52 | − 24 | − 33 | − 59 |

G
| 57 | 55 | 65 | 35 | 77 | 88 |
| − 37 | − 18 | − 28 | − 30 | − 45 | − 86 |

H
| 83 | 98 | 97 | 50 | 88 | 83 |
| − 29 | − 58 | − 24 | − 35 | − 35 | − 78 |

I
| 75 | 26 | 85 | 57 | 97 | 46 |
| − 69 | − 10 | − 69 | − 51 | − 91 | − 22 |

J
| 96 | 40 | 72 | 57 | 80 | 82 |
| − 22 | − 10 | − 39 | − 10 | − 24 | − 61 |

30

60

A
| 68 | 36 | 75 | 69 | 67 | 34 |
| − 43 | − 26 | − 74 | − 46 | − 36 | − 15 |

B
| 75 | 70 | 51 | 48 | 98 | 24 |
| − 57 | − 15 | − 11 | − 20 | − 31 | − 23 |

C
| 56 | 73 | 77 | 99 | 44 | 69 |
| − 46 | − 30 | − 39 | − 46 | − 20 | − 53 |

D
| 77 | 78 | 75 | 53 | 62 | 27 |
| − 10 | − 60 | − 37 | − 32 | − 23 | − 26 |

E
| 67 | 25 | 53 | 76 | 32 | 99 |
| − 49 | − 14 | − 37 | − 60 | − 28 | − 92 |

F
| 90 | 97 | 79 | 91 | 44 | 44 |
| − 80 | − 91 | − 67 | − 24 | − 34 | − 42 |

G
| 94 | 75 | 94 | 21 | 97 | 54 |
| − 13 | − 22 | − 72 | − 17 | − 42 | − 45 |

H
| 95 | 93 | 98 | 89 | 74 | 88 |
| − 43 | − 68 | − 61 | − 36 | − 46 | − 66 |

I
| 35 | 93 | 78 | 37 | 85 | 54 |
| − 30 | − 24 | − 21 | − 25 | − 82 | − 49 |

J
| 62 | 84 | 29 | 76 | 89 | 84 |
| − 32 | − 66 | − 13 | − 49 | − 17 | − 78 |

A

77	78	73	17	78	88
− 56	− 59	− 50	− 16	− 45	− 25

B

78	85	47	93	55	50
− 53	− 30	− 47	− 54	− 30	− 26

C

95	74	37	87	92	86
− 74	− 28	− 31	− 63	− 22	− 46

D

89	33	91	82	50	49
− 24	− 29	− 17	− 36	− 46	− 10

E

35	47	72	35	53	82
− 12	− 45	− 60	− 26	− 30	− 62

F

68	84	85	79	72	66
− 36	− 41	− 41	− 65	− 30	− 36

G

96	38	40	54	77	68
− 40	− 18	− 38	− 19	− 60	− 26

H

60	79	79	53	44	58
− 45	− 44	− 63	− 20	− 16	− 21

I

65	76	81	72	84	37
− 10	− 52	− 35	− 52	− 40	− 15

J

53	62	99	80	88	89
− 23	− 36	− 96	− 40	− 53	− 28

30

60

A	87 − 78	69 − 54	74 − 52	70 − 26	16 − 14	88 − 76
B	37 − 26	65 − 14	31 − 24	47 − 45	50 − 33	51 − 23
C	26 − 17	55 − 46	75 − 42	48 − 11	18 − 15	45 − 15
D	66 − 20	61 − 46	62 − 56	83 − 38	90 − 64	61 − 22
E	81 − 11	41 − 27	71 − 59	45 − 24	97 − 30	98 − 49
F	57 − 25	75 − 43	86 − 61	88 − 44	99 − 94	41 − 25
G	92 − 12	96 − 40	99 − 47	97 − 50	58 − 54	69 − 65
H	72 − 52	62 − 11	51 − 33	89 − 13	31 − 12	89 − 19
I	38 − 16	87 − 76	88 − 23	43 − 32	96 − 66	66 − 19
J	85 − 47	88 − 16	73 − 62	36 − 21	52 − 13	37 − 26

30

60

A
| 95 | 83 | 87 | 63 | 43 | 61 |
| − 63 | − 25 | − 61 | − 57 | − 27 | − 26 |

B
| 89 | 65 | 92 | 71 | 88 | 95 |
| − 37 | − 57 | − 75 | − 66 | − 69 | − 58 |

C
| 33 | 79 | 72 | 51 | 97 | 77 |
| − 20 | − 72 | − 15 | − 35 | − 58 | − 45 |

D
| 73 | 42 | 17 | 64 | 52 | 44 |
| − 21 | − 28 | − 15 | − 53 | − 45 | − 23 |

E
| 37 | 40 | 44 | 34 | 77 | 91 |
| − 36 | − 24 | − 27 | − 18 | − 65 | − 16 |

F
| 60 | 93 | 15 | 51 | 73 | 63 |
| − 36 | − 48 | − 14 | − 43 | − 51 | − 23 |

G
| 33 | 23 | 25 | 74 | 93 | 95 |
| − 18 | − 23 | − 11 | − 27 | − 29 | − 12 |

H
| 49 | 91 | 93 | 77 | 98 | 85 |
| − 30 | − 50 | − 46 | − 23 | − 96 | − 68 |

I
| 65 | 89 | 75 | 97 | 96 | 53 |
| − 21 | − 44 | − 56 | − 31 | − 58 | − 38 |

J
| 73 | 60 | 81 | 53 | 78 | 82 |
| − 66 | − 54 | − 19 | − 23 | − 45 | − 44 |

A
$$94 - 78$$ $$84 - 44$$ $$92 - 27$$ $$44 - 24$$ $$69 - 64$$ $$98 - 50$$

B
$$59 - 17$$ $$40 - 22$$ $$46 - 24$$ $$85 - 70$$ $$17 - 10$$ $$56 - 13$$

C
$$97 - 44$$ $$45 - 20$$ $$92 - 84$$ $$90 - 49$$ $$92 - 13$$ $$96 - 76$$

D
$$99 - 77$$ $$33 - 19$$ $$82 - 73$$ $$29 - 21$$ $$96 - 58$$ $$48 - 23$$

E
$$97 - 47$$ $$86 - 57$$ $$51 - 45$$ $$61 - 54$$ $$97 - 28$$ $$48 - 14$$

F
$$37 - 15$$ $$66 - 55$$ $$88 - 76$$ $$74 - 68$$ $$78 - 68$$ $$26 - 21$$

G
$$73 - 38$$ $$61 - 57$$ $$46 - 11$$ $$92 - 20$$ $$88 - 11$$ $$88 - 44$$

H
$$70 - 65$$ $$96 - 20$$ $$84 - 56$$ $$87 - 83$$ $$54 - 49$$ $$70 - 36$$

I
$$37 - 11$$ $$30 - 30$$ $$93 - 27$$ $$70 - 36$$ $$49 - 20$$ $$41 - 29$$

J
$$91 - 45$$ $$91 - 60$$ $$51 - 50$$ $$92 - 90$$ $$73 - 66$$ $$64 - 36$$

A
| 95 | 82 | 55 | 79 | 57 | 97 |
| − 86 | − 22 | − 22 | − 32 | − 43 | − 96 |

B
| 69 | 55 | 70 | 37 | 78 | 51 |
| − 12 | − 36 | − 35 | − 14 | − 68 | − 31 |

C
| 83 | 90 | 54 | 51 | 43 | 75 |
| − 23 | − 49 | − 17 | − 41 | − 36 | − 52 |

D
| 89 | 66 | 84 | 92 | 96 | 77 |
| − 69 | − 45 | − 22 | − 53 | − 63 | − 15 |

E
| 31 | 83 | 56 | 73 | 99 | 66 |
| − 29 | − 82 | − 32 | − 12 | − 42 | − 27 |

F
| 81 | 77 | 24 | 91 | 50 | 83 |
| − 64 | − 16 | − 22 | − 43 | − 29 | − 77 |

G
| 91 | 99 | 35 | 87 | 78 | 82 |
| − 88 | − 89 | − 30 | − 60 | − 16 | − 65 |

H
| 95 | 73 | 85 | 89 | 66 | 65 |
| − 86 | − 26 | − 16 | − 64 | − 65 | − 28 |

I
| 92 | 86 | 97 | 79 | 93 | 99 |
| − 59 | − 46 | − 96 | − 50 | − 49 | − 15 |

J
| 83 | 86 | 41 | 80 | 85 | 75 |
| − 74 | − 56 | − 21 | − 59 | − 17 | − 74 |

30

60

A	66 − 23	90 − 66	40 − 20	97 − 20	78 − 23	75 − 47
B	88 − 79	91 − 74	83 − 32	78 − 70	89 − 61	87 − 20
C	90 − 76	92 − 60	87 − 45	89 − 63	65 − 49	34 − 13
D	59 − 14	13 − 10	39 − 23	63 − 21	96 − 82	88 − 15
E	82 − 61	95 − 88	80 − 57	78 − 41	88 − 49	84 − 44
F	89 − 36	32 − 18	93 − 28	61 − 51	53 − 48	87 − 73
G	98 − 90	29 − 14	92 − 79	90 − 12	43 − 39	96 − 84
H	82 − 29	94 − 37	23 − 17	81 − 52	92 − 79	67 − 30
I	67 − 46	91 − 15	87 − 80	98 − 14	48 − 17	76 − 48
J	94 − 81	90 − 18	86 − 60	80 − 21	82 − 71	66 − 48

30

60

A	92 − 20	88 − 53	98 − 63	98 − 70	47 − 30	88 − 65
B	74 − 50	79 − 38	71 − 55	87 − 67	55 − 34	58 − 33
C	87 − 85	82 − 48	98 − 46	97 − 41	48 − 15	24 − 21
D	75 − 35	38 − 23	66 − 20	66 − 33	88 − 41	64 − 40
E	61 − 38	21 − 15	93 − 89	94 − 29	86 − 26	50 − 23
F	81 − 44	45 − 36	94 − 19	68 − 60	19 − 16	49 − 11
G	85 − 70	98 − 51	47 − 12	80 − 21	78 − 55	89 − 85
H	57 − 22	91 − 79	66 − 56	55 − 46	45 − 21	40 − 15
I	69 − 43	87 − 32	94 − 60	79 − 68	98 − 83	96 − 15
J	98 − 41	95 − 10	72 − 49	82 − 48	48 − 33	33 − 26

30

60

A

97	85	99	98	54	94
− 23	− 59	− 21	− 60	− 20	− 75

B

91	23	47	51	76	91
− 44	− 16	− 25	− 17	− 61	− 41

C

43	87	64	35	77	36
− 42	− 50	− 17	− 18	− 70	− 16

D

51	57	66	68	85	61
− 45	− 18	− 14	− 54	− 76	− 16

E

60	78	92	85	85	86
− 12	− 17	− 42	− 82	− 67	− 35

F

42	87	92	60	96	91
− 25	− 35	− 78	− 33	− 31	− 83

G

62	27	25	84	98	94
− 55	− 26	− 16	− 31	− 14	− 44

H

52	27	80	47	51	84
− 33	− 15	− 30	− 20	− 15	− 13

I

30	46	88	55	92	78
− 24	− 28	− 17	− 10	− 60	− 18

J

73	90	77	64	61	92
− 52	− 26	− 71	− 56	− 26	− 19

A	83 − 22	54 − 51	75 − 50	91 − 10	73 − 12	99 − 22
B	63 − 34	58 − 10	90 − 67	93 − 52	71 − 14	67 − 47
C	80 − 68	82 − 61	19 − 16	83 − 32	45 − 35	81 − 41
D	88 − 59	74 − 71	41 − 29	68 − 31	73 − 27	99 − 36
E	78 − 28	39 − 21	83 − 29	72 − 14	47 − 30	65 − 48
F	71 − 22	99 − 92	39 − 23	78 − 74	71 − 35	37 − 26
G	36 − 18	68 − 46	24 − 18	69 − 55	49 − 31	93 − 41
H	95 − 68	26 − 10	69 − 30	73 − 32	90 − 29	77 − 36
I	93 − 29	67 − 12	76 − 36	62 − 61	32 − 14	38 − 26
J	81 − 59	81 − 59	54 − 43	16 − 11	85 − 29	32 − 26

A	59 − 17	88 − 20	56 − 14	71 − 11	81 − 52	88 − 17
B	85 − 77	98 − 90	84 − 39	82 − 18	92 − 41	84 − 73
C	83 − 12	79 − 36	73 − 32	87 − 44	83 − 24	89 − 64
D	79 − 43	68 − 56	96 − 50	86 − 68	82 − 72	70 − 68
E	61 − 35	89 − 36	83 − 26	65 − 48	99 − 74	69 − 60
F	74 − 63	88 − 82	90 − 65	93 − 14	83 − 82	89 − 25
G	57 − 23	65 − 64	91 − 85	80 − 78	31 − 14	93 − 70
H	74 − 46	89 − 23	36 − 33	75 − 55	97 − 80	49 − 31
I	22 − 19	55 − 30	65 − 11	73 − 58	67 − 10	54 − 47
J	36 − 18	88 − 22	37 − 18	63 − 42	70 − 16	34 − 14

30

60

A
54	72	83	70	81	51
− 36	− 54	− 70	− 32	− 53	− 23

B
85	88	90	99	81	99
− 34	− 67	− 57	− 80	− 25	− 67

C
60	29	77	43	79	71
− 32	− 14	− 73	− 17	− 56	− 48

D
17	71	77	86	68	63
− 12	− 55	− 67	− 54	− 18	− 18

E
72	76	88	72	59	36
− 22	− 29	− 32	− 29	− 49	− 29

F
79	64	19	92	48	93
− 73	− 34	− 14	− 78	− 15	− 92

G
87	45	72	64	66	60
− 38	− 20	− 26	− 17	− 49	− 30

H
92	59	55	81	44	32
− 67	− 50	− 21	− 39	− 32	− 17

I
65	97	82	77	27	69
− 25	− 31	− 63	− 51	− 15	− 33

J
66	33	49	14	83	75
− 31	− 15	− 10	− 10	− 74	− 35

A	93 − 74	79 − 38	93 − 64	24 − 16	82 − 82	39 − 36
B	77 − 36	71 − 43	73 − 31	83 − 40	35 − 29	44 − 32
C	27 − 12	44 − 34	66 − 51	88 − 45	51 − 39	62 − 21
D	85 − 61	52 − 32	70 − 53	95 − 68	39 − 21	60 − 42
E	61 − 19	51 − 28	89 − 25	59 − 31	82 − 79	94 − 29
F	60 − 37	96 − 78	92 − 81	87 − 64	89 − 19	70 − 38
G	95 − 33	67 − 25	95 − 77	87 − 55	33 − 17	28 − 10
H	71 − 23	93 − 57	87 − 47	89 − 16	42 − 21	72 − 33
I	65 − 36	76 − 48	99 − 15	47 − 24	92 − 11	69 − 29
J	73 − 47	72 − 69	59 − 26	53 − 49	79 − 79	34 − 31

A

32	52	28	98	87	94
− 26	− 17	− 14	− 41	− 19	− 69

B

42	81	25	81	30	82
− 33	− 56	− 14	− 75	− 10	− 29

C

58	85	95	60	86	89
− 27	− 36	− 72	− 21	− 14	− 27

D

97	81	86	99	97	60
− 30	− 32	− 67	− 88	− 96	− 26

E

80	77	36	63	43	50
− 34	− 59	− 21	− 39	− 42	− 47

F

49	57	72	96	58	98
− 32	− 45	− 61	− 91	− 33	− 57

G

58	98	83	70	96	83
− 33	− 93	− 20	− 42	− 17	− 53

H

41	89	81	97	98	74
− 38	− 36	− 54	− 37	− 49	− 13

I

87	61	46	84	95	57
− 53	− 39	− 32	− 19	− 21	− 10

J

50	54	49	46	45	31
− 34	− 19	− 45	− 25	− 32	− 23

A
97	41	99	89	84	43
− 88	− 39	− 22	− 49	− 59	− 31

B
73	86	50	57	50	92
− 52	− 72	− 29	− 12	− 38	− 35

C
85	44	84	78	63	88
− 33	− 40	− 26	− 46	− 59	− 46

D
92	80	85	81	95	86
− 72	− 12	− 76	− 75	− 75	− 40

E
87	84	32	92	31	94
− 34	− 19	− 32	− 46	− 26	− 78

F
52	81	82	84	84	45
− 34	− 46	− 60	− 43	− 81	− 29

G
51	83	89	39	92	92
− 48	− 15	− 88	− 16	− 47	− 51

H
54	31	56	62	98	42
− 15	− 18	− 49	− 40	− 65	− 18

I
98	82	26	66	87	87
− 73	− 58	− 26	− 29	− 15	− 35

J
77	89	44	53	73	67
− 13	− 33	− 41	− 24	− 30	− 17

30

60

A

| 36
− 28 | 94
− 35 | 71
− 44 | 36
− 17 | 69
− 67 | 76
− 17 |

B

| 93
− 53 | 71
− 41 | 52
− 24 | 65
− 24 | 63
− 38 | 90
− 74 |

C

| 90
− 31 | 59
− 30 | 94
− 36 | 85
− 77 | 82
− 41 | 41
− 23 |

D

| 40
− 27 | 56
− 42 | 81
− 74 | 39
− 37 | 87
− 52 | 59
− 23 |

E

| 30
− 14 | 88
− 55 | 89
− 81 | 30
− 30 | 85
− 13 | 60
− 59 |

F

| 34
− 28 | 25
− 13 | 96
− 29 | 97
− 68 | 72
− 43 | 44
− 43 |

G

| 55
− 43 | 84
− 82 | 68
− 26 | 98
− 80 | 89
− 49 | 95
− 23 |

H

| 71
− 43 | 81
− 12 | 99
− 88 | 80
− 44 | 95
− 34 | 88
− 42 |

I

| 86
− 24 | 88
− 57 | 90
− 26 | 45
− 18 | 29
− 16 | 98
− 42 |

J

| 42
− 25 | 74
− 67 | 89
− 65 | 58
− 16 | 85
− 82 | 76
− 48 |

A
| 48
− 39 | 90
− 85 | 94
− 40 | 90
− 89 | 93
− 55 | 56
− 11 |

B
| 25
− 11 | 37
− 26 | 45
− 31 | 67
− 15 | 96
− 86 | 66
− 45 |

C
| 27
− 24 | 40
− 35 | 44
− 43 | 78
− 23 | 81
− 37 | 77
− 71 |

D
| 94
− 86 | 27
− 17 | 96
− 91 | 58
− 28 | 28
− 26 | 79
− 48 |

E
| 75
− 74 | 99
− 79 | 82
− 70 | 77
− 47 | 59
− 21 | 93
− 22 |

F
| 44
− 39 | 81
− 20 | 31
− 16 | 54
− 48 | 95
− 15 | 52
− 52 |

G
| 91
− 88 | 79
− 78 | 88
− 74 | 72
− 64 | 59
− 19 | 77
− 64 |

H
| 83
− 72 | 70
− 28 | 62
− 56 | 69
− 38 | 85
− 53 | 57
− 27 |

I
| 48
− 47 | 29
− 14 | 32
− 27 | 52
− 14 | 90
− 51 | 99
− 86 |

J
| 86
− 77 | 91
− 91 | 42
− 14 | 79
− 59 | 75
− 44 | 47
− 24 |

A	22 - 20	71 - 10	95 - 41	79 - 45	47 - 15	78 - 47
B	81 - 58	39 - 10	37 - 21	49 - 35	22 - 16	32 - 18
C	95 - 29	67 - 16	73 - 43	86 - 36	71 - 38	82 - 63
D	36 - 28	82 - 45	33 - 22	96 - 53	96 - 25	81 - 75
E	61 - 56	78 - 70	88 - 56	75 - 15	69 - 31	45 - 21
F	99 - 52	49 - 20	76 - 41	98 - 84	53 - 43	99 - 31
G	29 - 26	55 - 27	40 - 15	65 - 33	95 - 18	73 - 72
H	85 - 16	51 - 47	60 - 31	83 - 76	70 - 30	69 - 61
I	80 - 15	40 - 11	87 - 56	78 - 55	52 - 21	70 - 20
J	75 - 12	83 - 68	86 - 64	91 - 51	51 - 37	64 - 27

A

37	32	42	24	29	84
− 24	− 11	− 21	− 21	− 26	− 41

B

95	37	75	69	70	84
− 84	− 15	− 22	− 63	− 52	− 27

C

18	62	77	76	34	69
− 13	− 54	− 27	− 62	− 26	− 11

D

99	94	60	64	67	77
− 53	− 65	− 59	− 22	− 35	− 39

E

38	27	20	85	47	87
− 23	− 10	− 19	− 74	− 45	− 52

F

74	57	61	74	76	58
− 73	− 43	− 24	− 24	− 46	− 16

G

83	71	77	78	33	50
− 18	− 63	− 56	− 53	− 12	− 23

H

68	76	44	83	89	98
− 62	− 29	− 14	− 66	− 39	− 92

I

79	71	45	96	63	87
− 51	− 56	− 34	− 49	− 57	− 42

J

59	45	66	97	74	60
− 19	− 16	− 63	− 76	− 15	− 10

A

89	80	19	88	81	89
− 15	− 31	− 17	− 56	− 74	− 53

B

98	23	64	90	68	52
− 70	− 21	− 21	− 61	− 23	− 21

C

67	39	65	68	50	45
− 65	− 39	− 30	− 58	− 29	− 33

D

84	47	62	36	97	42
− 18	− 37	− 34	− 26	− 65	− 28

E

98	88	50	86	65	69
− 38	− 44	− 44	− 20	− 16	− 20

F

40	98	67	96	26	32
− 34	− 46	− 42	− 27	− 22	− 25

G

73	88	69	82	55	76
− 45	− 29	− 33	− 26	− 39	− 44

H

83	48	78	58	26	39
− 51	− 21	− 30	− 40	− 12	− 22

I

54	42	92	74	79	70
− 36	− 11	− 48	− 47	− 44	− 69

J

82	61	57	80	35	68
− 80	− 12	− 52	− 64	− 10	− 46

A
| 84 | 57 | 64 | 82 | 64 | 49 |
| − 10 | − 17 | − 30 | − 36 | − 36 | − 21 |

B
| 57 | 85 | 80 | 68 | 83 | 98 |
| − 44 | − 28 | − 36 | − 67 | − 65 | − 85 |

C
| 22 | 93 | 48 | 66 | 40 | 83 |
| − 13 | − 50 | − 39 | − 32 | − 23 | − 27 |

D
| 86 | 79 | 79 | 46 | 89 | 96 |
| − 22 | − 78 | − 78 | − 16 | − 76 | − 90 |

E
| 34 | 91 | 29 | 48 | 87 | 51 |
| − 27 | − 34 | − 15 | − 13 | − 70 | − 30 |

F
| 66 | 61 | 98 | 87 | 67 | 80 |
| − 19 | − 19 | − 58 | − 48 | − 12 | − 61 |

G
| 81 | 81 | 87 | 32 | 99 | 61 |
| − 23 | − 54 | − 39 | − 21 | − 25 | − 16 |

H
| 91 | 50 | 34 | 53 | 45 | 99 |
| − 58 | − 13 | − 25 | − 50 | − 19 | − 98 |

I
| 32 | 44 | 80 | 90 | 86 | 91 |
| − 23 | − 42 | − 50 | − 20 | − 34 | − 91 |

J
| 95 | 87 | 77 | 76 | 56 | 90 |
| − 83 | − 65 | − 40 | − 37 | − 31 | − 34 |

A
| 79 | 44 | 94 | 91 | 88 | 55 |
| − 45 | − 42 | − 61 | − 14 | − 50 | − 22 |

B
| 38 | 80 | 56 | 59 | 91 | 33 |
| − 36 | − 21 | − 36 | − 20 | − 40 | − 30 |

C
| 74 | 80 | 90 | 58 | 90 | 36 |
| − 19 | − 39 | − 11 | − 26 | − 67 | − 17 |

D
| 89 | 89 | 97 | 87 | 32 | 54 |
| − 69 | − 81 | − 25 | − 73 | − 30 | − 44 |

E
| 41 | 70 | 78 | 99 | 84 | 41 |
| − 11 | − 42 | − 41 | − 26 | − 74 | − 36 |

F
| 48 | 72 | 29 | 29 | 46 | 73 |
| − 22 | − 68 | − 14 | − 26 | − 43 | − 70 |

G
| 60 | 84 | 80 | 73 | 92 | 16 |
| − 24 | − 29 | − 18 | − 47 | − 79 | − 13 |

H
| 89 | 91 | 98 | 78 | 89 | 51 |
| − 54 | − 10 | − 47 | − 34 | − 32 | − 34 |

I
| 73 | 65 | 31 | 79 | 92 | 60 |
| − 48 | − 13 | − 11 | − 43 | − 72 | − 41 |

J
| 85 | 82 | 75 | 58 | 77 | 86 |
| − 38 | − 47 | − 51 | − 27 | − 27 | − 63 |

A	79 − 24	81 − 22	75 − 29	63 − 44	62 − 29	64 − 53
B	93 − 82	93 − 52	57 − 29	68 − 49	56 − 48	94 − 85
C	75 − 62	57 − 31	48 − 30	87 − 26	62 − 23	73 − 30
D	68 − 60	56 − 50	73 − 39	73 − 24	80 − 52	87 − 86
E	56 − 19	68 − 13	38 − 10	71 − 32	88 − 44	74 − 19
F	70 − 16	82 − 47	71 − 14	73 − 10	94 − 38	82 − 28
G	65 − 23	47 − 37	64 − 17	96 − 42	51 − 36	55 − 14
H	60 − 25	84 − 23	92 − 21	19 − 10	65 − 35	80 − 14
I	63 − 49	78 − 28	93 − 48	60 − 28	61 − 36	78 − 69
J	88 − 12	73 − 34	24 − 22	20 − 13	81 − 16	97 − 96

30

60

A

73	71	95	84	44	86
− 17	− 15	− 30	− 12	− 39	− 62

B

78	81	85	90	42	83
− 34	− 55	− 45	− 85	− 15	− 15

C

98	87	47	58	84	35
− 98	− 84	− 31	− 43	− 50	− 29

D

95	93	68	68	86	37
− 40	− 78	− 17	− 51	− 35	− 36

E

66	53	92	67	92	84
− 54	− 11	− 60	− 59	− 17	− 75

F

74	24	99	86	81	95
− 28	− 21	− 63	− 33	− 49	− 31

G

37	94	83	95	70	28
− 21	− 56	− 42	− 37	− 22	− 21

H

58	14	48	79	61	90
− 45	− 10	− 21	− 57	− 40	− 19

I

50	55	80	96	33	80
− 23	− 29	− 79	− 39	− 26	− 72

J

70	65	65	99	86	76
− 45	− 52	− 48	− 25	− 30	− 74

Day: 102
Date:
Score: /60
Name:
Time: :
Rating: ☆☆☆☆☆

A	29 − 13	62 − 36	43 − 17	63 − 24	93 − 54	78 − 39
B	65 − 53	92 − 17	48 − 31	54 − 37	93 − 41	32 − 31
C	95 − 60	76 − 31	35 − 20	83 − 64	63 − 20	35 − 29
D	38 − 37	77 − 68	71 − 37	50 − 13	88 − 32	92 − 41
E	69 − 47	87 − 62	33 − 27	97 − 26	95 − 33	79 − 21
F	71 − 12	63 − 46	64 − 19	87 − 86	20 − 13	92 − 19
G	52 − 38	59 − 53	42 − 19	80 − 67	61 − 51	58 − 13
H	51 − 37	74 − 60	64 − 22	96 − 22	82 − 23	61 − 38
I	83 − 41	26 − 10	14 − 10	56 − 34	63 − 22	32 − 28
J	88 − 84	66 − 31	89 − 12	96 − 16	75 − 26	85 − 77

30

60

A	88 − 14	97 − 33	95 − 59	97 − 79	36 − 15	97 − 45
B	92 − 30	80 − 55	78 − 43	83 − 61	34 − 33	99 − 67
C	41 − 18	63 − 35	97 − 56	49 − 47	73 − 44	53 − 47
D	68 − 31	62 − 56	71 − 30	82 − 81	82 − 52	21 − 11
E	22 − 15	87 − 81	85 − 75	97 − 58	67 − 41	87 − 67
F	59 − 44	86 − 45	93 − 72	76 − 60	53 − 35	55 − 20
G	66 − 62	77 − 19	87 − 32	80 − 78	91 − 75	85 − 84
H	95 − 16	91 − 73	95 − 17	67 − 51	73 − 32	95 − 78
I	65 − 10	65 − 14	70 − 41	44 − 43	79 − 39	25 − 22
J	92 − 63	78 − 45	81 − 49	71 − 11	94 − 69	29 − 18

30

60

A

73	41	73	76	96	66
− 28	− 33	− 22	− 45	− 90	− 11

B

81	73	78	45	83	74
− 64	− 69	− 10	− 12	− 72	− 14

C

74	45	39	72	91	20
− 37	− 15	− 16	− 36	− 71	− 14

D

95	95	93	31	63	87
− 54	− 26	− 50	− 23	− 18	− 86

E

90	84	99	94	86	82
− 24	− 45	− 58	− 75	− 71	− 44

F

84	98	93	86	39	81
− 35	− 27	− 67	− 15	− 24	− 71

G

74	20	84	97	96	87
− 53	− 15	− 41	− 42	− 31	− 22

H

60	70	36	93	47	83
− 43	− 51	− 36	− 27	− 28	− 67

I

92	98	53	80	99	85
− 23	− 17	− 40	− 49	− 79	− 14

J

84	59	59	66	83	90
− 81	− 58	− 49	− 22	− 64	− 79

A
| 54 − 10 | 92 − 65 | 65 − 53 | 65 − 22 | 50 − 23 | 84 − 67 |

B
| 36 − 18 | 35 − 12 | 36 − 20 | 30 − 16 | 49 − 36 | 99 − 99 |

C
| 96 − 89 | 68 − 60 | 81 − 69 | 55 − 20 | 83 − 76 | 92 − 61 |

D
| 94 − 60 | 83 − 12 | 87 − 58 | 66 − 40 | 96 − 87 | 69 − 19 |

E
| 74 − 26 | 93 − 12 | 87 − 50 | 37 − 27 | 89 − 38 | 54 − 38 |

F
| 68 − 11 | 60 − 29 | 53 − 21 | 64 − 44 | 98 − 26 | 45 − 31 |

G
| 96 − 64 | 25 − 18 | 22 − 16 | 93 − 24 | 98 − 42 | 59 − 33 |

H
| 92 − 58 | 98 − 15 | 97 − 85 | 60 − 47 | 70 − 68 | 73 − 72 |

I
| 29 − 23 | 90 − 16 | 77 − 47 | 90 − 12 | 92 − 53 | 78 − 15 |

J
| 86 − 70 | 84 − 59 | 79 − 61 | 90 − 44 | 65 − 38 | 99 − 22 |

A
| 57 | 92 | 41 | 70 | 73 | 46 |
| − 32 | − 87 | − 12 | − 28 | − 67 | − 44 |

B
| 66 | 52 | 38 | 91 | 63 | 96 |
| − 12 | − 23 | − 29 | − 49 | − 13 | − 70 |

C
| 72 | 58 | 58 | 95 | 90 | 80 |
| − 22 | − 52 | − 53 | − 76 | − 12 | − 42 |

D
| 54 | 99 | 80 | 89 | 87 | 59 |
| − 47 | − 60 | − 46 | − 19 | − 80 | − 29 |

E
| 83 | 99 | 87 | 92 | 86 | 40 |
| − 52 | − 63 | − 16 | − 76 | − 83 | − 13 |

F
| 50 | 65 | 52 | 77 | 84 | 55 |
| − 18 | − 43 | − 36 | − 57 | − 68 | − 14 |

G
| 46 | 93 | 32 | 86 | 78 | 90 |
| − 33 | − 57 | − 26 | − 64 | − 23 | − 57 |

H
| 92 | 98 | 68 | 92 | 84 | 83 |
| − 53 | − 47 | − 15 | − 85 | − 20 | − 54 |

I
| 85 | 36 | 80 | 47 | 64 | 32 |
| − 47 | − 21 | − 13 | − 38 | − 46 | − 23 |

J
| 22 | 54 | 58 | 61 | 79 | 66 |
| − 18 | − 47 | − 18 | − 37 | − 49 | − 65 |

30

60

A
| 73 | 76 | 88 | 91 | 46 | 43 |
| − 26 | − 40 | − 67 | − 74 | − 42 | − 24 |

B
| 88 | 53 | 49 | 90 | 39 | 84 |
| − 37 | − 19 | − 14 | − 84 | − 28 | − 15 |

C
| 98 | 66 | 42 | 68 | 93 | 86 |
| − 24 | − 31 | − 34 | − 10 | − 90 | − 76 |

D
| 83 | 81 | 52 | 68 | 72 | 92 |
| − 71 | − 66 | − 37 | − 17 | − 66 | − 47 |

E
| 96 | 93 | 89 | 49 | 33 | 89 |
| − 62 | − 41 | − 81 | − 27 | − 19 | − 45 |

F
| 73 | 33 | 51 | 44 | 42 | 96 |
| − 11 | − 16 | − 42 | − 15 | − 42 | − 19 |

G
| 97 | 92 | 82 | 45 | 77 | 46 |
| − 91 | − 23 | − 15 | − 24 | − 22 | − 35 |

H
| 37 | 52 | 82 | 95 | 31 | 15 |
| − 36 | − 40 | − 54 | − 79 | − 18 | − 13 |

I
| 83 | 61 | 69 | 53 | 61 | 70 |
| − 18 | − 47 | − 25 | − 47 | − 13 | − 24 |

J
| 81 | 83 | 97 | 70 | 47 | 31 |
| − 70 | − 32 | − 46 | − 28 | − 18 | − 23 |

A
$$73 - 18$$ $$64 - 48$$ $$36 - 17$$ $$82 - 56$$ $$91 - 63$$ $$96 - 91$$

B
$$90 - 86$$ $$65 - 45$$ $$65 - 48$$ $$69 - 62$$ $$68 - 43$$ $$78 - 73$$

C
$$91 - 48$$ $$65 - 31$$ $$36 - 25$$ $$95 - 60$$ $$68 - 33$$ $$50 - 19$$

D
$$79 - 53$$ $$92 - 62$$ $$83 - 79$$ $$70 - 62$$ $$64 - 28$$ $$73 - 61$$

E
$$60 - 46$$ $$90 - 11$$ $$70 - 63$$ $$79 - 26$$ $$56 - 36$$ $$12 - 11$$

F
$$80 - 57$$ $$74 - 62$$ $$75 - 50$$ $$86 - 68$$ $$91 - 68$$ $$57 - 51$$

G
$$64 - 38$$ $$40 - 13$$ $$96 - 37$$ $$29 - 23$$ $$51 - 15$$ $$53 - 16$$

H
$$91 - 30$$ $$36 - 33$$ $$86 - 54$$ $$57 - 20$$ $$56 - 48$$ $$50 - 35$$

I
$$45 - 18$$ $$96 - 38$$ $$55 - 20$$ $$86 - 56$$ $$88 - 88$$ $$26 - 16$$

J
$$70 - 36$$ $$60 - 18$$ $$41 - 28$$ $$36 - 26$$ $$98 - 67$$ $$77 - 25$$

30

60

A	83 − 49	81 − 17	86 − 19	82 − 70	29 − 19	79 − 53
B	57 − 52	46 − 31	80 − 64	60 − 24	58 − 14	77 − 23
C	58 − 20	72 − 13	94 − 72	81 − 34	83 − 61	33 − 10
D	35 − 28	67 − 32	39 − 33	98 − 55	80 − 59	80 − 39
E	86 − 62	97 − 84	91 − 52	52 − 48	64 − 31	76 − 63
F	47 − 41	56 − 16	68 − 49	78 − 26	77 − 39	89 − 69
G	86 − 42	84 − 22	61 − 58	91 − 83	63 − 48	98 − 73
H	63 − 27	86 − 51	67 − 30	54 − 34	76 − 54	36 − 33
I	69 − 27	83 − 18	74 − 24	59 − 16	55 − 21	54 − 44
J	97 − 85	61 − 52	72 − 50	76 − 36	47 − 41	91 − 85

30

60

A
| 83 | 89 | 84 | 56 | 21 | 95 |
| - 80 | - 73 | - 54 | - 32 | - 12 | - 17 |

B
| 60 | 71 | 43 | 75 | 50 | 86 |
| - 43 | - 63 | - 24 | - 11 | - 44 | - 82 |

C
| 94 | 19 | 47 | 57 | 65 | 68 |
| - 27 | - 16 | - 35 | - 48 | - 12 | - 47 |

D
| 85 | 90 | 49 | 98 | 45 | 82 |
| - 83 | - 61 | - 47 | - 78 | - 41 | - 47 |

E
| 69 | 60 | 79 | 53 | 68 | 80 |
| - 46 | - 45 | - 71 | - 40 | - 11 | - 31 |

F
| 34 | 24 | 52 | 47 | 59 | 48 |
| - 18 | - 19 | - 28 | - 42 | - 26 | - 14 |

G
| 64 | 51 | 73 | 69 | 80 | 87 |
| - 34 | - 40 | - 35 | - 65 | - 13 | - 32 |

H
| 98 | 43 | 73 | 68 | 64 | 48 |
| - 95 | - 26 | - 65 | - 53 | - 12 | - 39 |

I
| 88 | 70 | 56 | 41 | 85 | 78 |
| - 62 | - 17 | - 43 | - 35 | - 51 | - 65 |

J
| 84 | 75 | 57 | 81 | 54 | 63 |
| - 54 | - 27 | - 22 | - 15 | - 47 | - 12 |

Bonus Sheets - Triple Digit Addition

A	258 + 738	457 + 437	687 + 859	666 + 906	150 + 247	831 + 410
B	940 + 532	970 + 110	838 + 484	178 + 163	705 + 513	956 + 182
C	775 + 506	692 + 990	613 + 948	228 + 599	953 + 505	127 + 152
D	437 + 559	538 + 801	617 + 895	440 + 132	807 + 514	231 + 875
E	210 + 459	914 + 178	536 + 945	267 + 162	935 + 704	786 + 308
F	298 + 321	851 + 328	215 + 780	827 + 874	379 + 877	659 + 634
G	834 + 518	852 + 733	275 + 378	926 + 394	986 + 216	888 + 357
H	660 + 355	121 + 613	261 + 865	997 + 525	496 + 786	705 + 770
I	782 + 680	354 + 835	313 + 899	523 + 830	869 + 923	432 + 697
J	706 + 294	284 + 719	954 + 545	167 + 944	939 + 660	528 + 811

Bonus Sheets - Triple Digit Subtraction

A	774 − 470	823 − 356	715 − 371	608 − 372	995 − 462	816 − 357
B	432 − 242	978 − 508	821 − 658	765 − 558	904 − 598	647 − 472
C	958 − 138	629 − 529	823 − 729	312 − 280	802 − 678	587 − 395
D	706 − 668	774 − 509	983 − 699	706 − 330	184 − 115	580 − 554
E	889 − 486	815 − 363	857 − 531	652 − 102	898 − 195	800 − 572
F	523 − 520	494 − 480	997 − 663	603 − 241	871 − 138	751 − 118
G	154 − 152	856 − 375	493 − 372	680 − 557	918 − 499	610 − 336
H	413 − 337	665 − 486	426 − 370	702 − 524	672 − 409	765 − 600
I	888 − 292	657 − 494	514 − 316	691 − 351	840 − 565	443 − 299
J	796 − 280	690 − 165	819 − 575	901 − 811	873 − 466	942 − 792

Addition Answer Key Sheet

DAY 1
- A 73, 161, 72, 26, 75, 105
- F 147, 128, 71, 139, 77, 154
- B 28, 128, 71, 129, 54, 95
- G 107, 120, 120, 77, 94, 102
- C 62, 100, 109, 86, 160, 73
- H 132, 63, 153, 129, 100, 130
- D 92, 145, 135, 154, 125, 118
- I 144, 98, 90, 110, 62, 115
- E 140, 116, 114, 66, 190, 76
- J 78, 112, 142, 113, 195, 119

DAY 2
- A 103, 101, 160, 165, 56, 108
- F 105, 114, 189, 124, 111, 122
- B 160, 107, 63, 181, 156, 71
- G 79, 146, 94, 90, 132, 68
- C 94, 102, 84, 135, 103, 52
- H 103, 118, 120, 134, 103, 142
- D 82, 155, 80, 114, 116, 162
- I 107, 167, 143, 43, 127, 170
- E 62, 156, 135, 89, 68, 103
- J 76, 73, 111, 91, 70, 49

DAY 3
- A 127, 125, 86, 73, 76, 89
- F 180, 115, 129, 40, 94, 104
- B 90, 49, 106, 94, 120, 41
- G 150, 91, 69, 127, 120, 88
- C 119, 48, 180, 168, 96, 135
- H 101, 101, 63, 60, 162, 171
- D 157, 55, 116, 104, 120, 83
- I 99, 34, 126, 104, 141, 124
- E 186, 92, 79, 168, 115, 39
- J 170, 111, 174, 178, 172, 112

DAY 4
- A 114, 135, 66, 84, 159, 160
- F 55, 100, 80, 78, 102, 112
- B 179, 78, 49, 105, 40, 79
- G 180, 118, 135, 188, 68, 104
- C 45, 138, 108, 44, 123, 136
- H 80, 65, 108, 108, 81, 122
- D 146, 44, 121, 94, 102, 138
- I 87, 142, 138, 30, 98, 47
- E 120, 106, 111, 112, 96, 90
- J 115, 156, 69, 172, 115, 51

DAY 5
- A 72, 128, 47, 133, 85, 65
- F 38, 173, 132, 110, 150, 71
- B 123, 64, 143, 71, 67, 75
- G 92, 155, 113, 84, 91, 126
- C 135, 68, 125, 194, 127, 140
- H 127, 123, 100, 88, 129, 128
- D 98, 168, 153, 177, 63, 99
- I 68, 154, 62, 110, 185, 77
- E 67, 144, 69, 123, 125, 88
- J 127, 105, 58, 97, 88, 98

DAY 6
- A 111, 114, 108, 160, 137, 156
- F 125, 122, 187, 95, 176, 61
- B 150, 120, 61, 139, 151, 41
- G 61, 67, 103, 94, 83, 84
- C 95, 103, 119, 186, 131, 167
- H 72, 101, 133, 116, 150, 143
- D 146, 186, 93, 55, 136, 74
- I 140, 129, 82, 103, 138, 95
- E 103, 117, 76, 144, 152, 101
- J 67, 109, 131, 99, 73, 90

DAY 7
- A 50, 80, 92, 141, 163, 124
- F 136, 102, 108, 159, 94, 83
- B 130, 54, 38, 172, 67, 79
- G 147, 109, 86, 163, 102, 102
- C 176, 134, 32, 101, 92, 64
- H 57, 89, 39, 92, 116, 92
- D 74, 91, 155, 133, 156, 83
- I 137, 116, 131, 45, 168, 121
- E 186, 101, 138, 37, 127, 123
- J 104, 33, 106, 162, 148, 96

DAY 8
- A 104, 75, 87, 129, 98, 129
- F 44, 163, 101, 102, 111, 39
- B 136, 138, 115, 80, 134, 134
- G 120, 115, 112, 44, 87, 98
- C 61, 109, 93, 107, 73, 63
- H 102, 54, 142, 112, 175, 124
- D 70, 67, 46, 120, 184, 130
- I 112, 28, 138, 188, 114, 111
- E 182, 114, 52, 61, 21, 68
- J 116, 38, 110, 63, 106, 85

DAY 9
- A 123, 131, 170, 126, 111, 133
- F 51, 112, 164, 83, 78, 106
- B 87, 186, 152, 152, 150, 121
- G 83, 65, 133, 107, 27, 117
- C 100, 108, 105, 87, 102, 122
- H 96, 152, 64, 116, 147, 155
- D 93, 135, 84, 179, 82, 100
- I 100, 75, 131, 165, 100, 113
- E 124, 88, 106, 165, 155, 107
- J 170, 76, 118, 139, 65, 143

DAY 10
- A 61, 95, 173, 46, 80, 89
- F 103, 134, 134, 149, 121, 122
- B 183, 107, 183, 117, 58, 90
- G 132, 163, 124, 78, 117, 124
- C 151, 53, 139, 78, 149, 150
- H 102, 132, 108, 107, 80, 68
- D 98, 95, 26, 137, 101, 47
- I 185, 81, 139, 56, 148, 142
- E 117, 52, 67, 39, 139, 110
- J 104, 88, 143, 150, 75, 118

DAY 11
- A 118, 174, 109, 102, 146, 122
- F 68, 120, 114, 82, 54, 151
- B 120, 157, 145, 81, 149, 166
- G 117, 79, 150, 82, 143, 113
- C 50, 81, 166, 38, 92, 149
- H 116, 131, 81, 93, 57, 123
- D 118, 120, 117, 94, 94, 170
- I 176, 145, 65, 111, 110, 195
- E 92, 111, 149, 125, 97, 100
- J 134, 78, 118, 121, 168, 85

DAY 12
- A 98, 108, 67, 142, 143, 156
- F 123, 111, 76, 38, 28, 128
- B 99, 124, 102, 97, 95, 120
- G 161, 101, 45, 170, 113, 115
- C 118, 155, 92, 168, 169, 107
- H 64, 108, 95, 74, 71, 66
- D 45, 115, 38, 108, 78, 71
- I 96, 84, 115, 82, 135, 95
- E 29, 110, 165, 136, 76, 51
- J 155, 120, 48, 102, 145, 85

DAY 13
- A 83, 95, 118, 60, 98, 52
- F 69, 148, 36, 133, 176, 88
- B 115, 123, 76, 178, 50, 69
- G 72, 124, 84, 57, 97, 143
- C 101, 67, 159, 137, 91, 76
- H 106, 183, 71, 43, 132, 182
- D 109, 147, 85, 62, 75, 117
- I 88, 49, 150, 94, 174, 112
- E 81, 55, 163, 156, 144, 34
- J 111, 92, 96, 147, 140, 152

DAY 14
- A 142, 191, 144, 165, 118, 110
- F 124, 133, 147, 47, 110, 50
- B 177, 110, 103, 107, 40, 125
- G 89, 96, 62, 113, 187, 94
- C 146, 76, 87, 31, 175, 103
- H 143, 148, 60, 110, 83, 103
- D 98, 38, 153, 67, 121, 68
- I 142, 159, 21, 38, 59, 74
- E 175, 90, 79, 136, 81, 136
- J 140, 45, 194, 63, 91, 170

DAY 15
- A 85, 122, 164, 120, 130, 110
- F 161, 151, 84, 106, 99, 69
- B 87, 114, 156, 105, 131, 96
- G 74, 82, 38, 97, 74, 119
- C 127, 124, 132, 174, 51, 181
- H 114, 138, 106, 159, 134, 131
- D 131, 74, 156, 88, 132, 146
- I 122, 137, 104, 159, 82, 68
- E 128, 126, 141, 144, 132, 112
- J 192, 36, 106, 71, 196, 74

DAY 16
- A 38, 104, 123, 126, 84, 118
- F 98, 138, 139, 180, 110, 68
- B 127, 110, 66, 57, 106, 87
- G 137, 106, 134, 63, 172, 110
- C 148, 181, 62, 54, 111, 151
- H 168, 147, 109, 75, 104, 115
- D 127, 128, 63, 113, 92, 156
- I 136, 132, 99, 82, 102, 77
- E 167, 109, 91, 95, 90, 114
- J 39, 153, 95, 71, 100, 129

DAY 17
- A 124, 106, 126, 156, 95, 173
- F 94, 125, 120, 116, 132, 91
- B 34, 142, 33, 65, 121, 58
- G 137, 155, 157, 44, 88, 98
- C 153, 110, 156, 49, 150, 125
- H 83, 107, 74, 130, 131, 172
- D 108, 56, 90, 121, 158, 131
- I 94, 132, 101, 157, 108, 179
- E 60, 147, 83, 113, 107, 115
- J 116, 101, 50, 104, 135, 174

DAY 18
- A 161, 90, 133, 92, 140, 107
- F 108, 119, 124, 113, 112, 134
- B 102, 47, 186, 97, 131, 139
- G 118, 94, 55, 105, 163, 96
- C 84, 98, 103, 86, 97, 151
- H 113, 70, 91, 83, 175, 136
- D 53, 62, 118, 89, 105, 99
- I 72, 176, 126, 84, 95, 67
- E 116, 81, 40, 131, 113, 140
- J 115, 181, 74, 146, 106, 53

DAY 19
- A 58, 76, 77, 138, 114, 107
- F 91, 78, 135, 95, 73, 110
- B 124, 149, 98, 119, 73, 159
- G 69, 124, 108, 136, 138, 102
- C 113, 110, 127, 44, 180, 118
- H 138, 94, 68, 72, 104, 169
- D 128, 158, 63, 91, 130, 154
- I 145, 129, 85, 120, 122, 120
- E 153, 107, 138, 112, 139, 110
- J 82, 114, 121, 98, 122, 138

DAY 20
- A 96, 123, 157, 89, 94, 160
- F 80, 146, 154, 173, 53, 133
- B 94, 124, 136, 88, 99, 76
- G 163, 127, 138, 149, 148, 71
- C 179, 92, 50, 72, 50, 153
- H 32, 97, 71, 114, 74, 62
- D 79, 122, 160, 84, 66, 111
- I 68, 141, 130, 105, 164, 126
- E 97, 70, 113, 103, 149, 63
- J 108, 32, 72, 57, 124, 27

DAY 21
- A 111, 63, 51, 89, 76, 117
- F 104, 69, 44, 101, 108, 100
- B 75, 100, 159, 165, 88, 81
- G 79, 176, 136, 186, 167, 33
- C 69, 87, 116, 143, 128, 175
- H 175, 111, 39, 115, 102, 28
- D 99, 115, 85, 91, 120, 88
- I 170, 115, 89, 103, 105, 187
- E 77, 163, 90, 148, 105, 112
- J 77, 118, 182, 67, 124, 132

DAY 22
- A 56, 143, 91, 116, 114, 181
- F 134, 170, 152, 122, 109, 124
- B 57, 80, 100, 87, 110, 63
- G 116, 96, 104, 101, 116, 132
- C 115, 144, 85, 49, 79, 114
- H 120, 110, 57, 169, 102, 112
- D 65, 123, 122, 167, 159, 107
- I 149, 138, 149, 92, 154, 120
- E 101, 175, 169, 104, 138, 91
- J 101, 55, 63, 118, 68, 24

DAY 23
- A 106, 91, 147, 142, 35, 39
- F 113, 124, 175, 89, 97, 49
- B 92, 157, 96, 132, 160, 61
- G 82, 93, 96, 82, 22, 85
- C 80, 83, 95, 140, 167, 88
- H 129, 87, 138, 65, 78, 145
- D 121, 137, 71, 162, 158, 190
- I 74, 120, 171, 85, 189, 125
- E 103, 83, 74, 102, 85, 169
- J 86, 91, 111, 67, 97, 95

DAY 24
- A 116, 185, 93, 78, 123, 93
- F 140, 54, 123, 107, 39, 161
- B 90, 152, 50, 101, 174, 50
- G 58, 162, 136, 171, 146, 119
- C 128, 159, 127, 153, 140, 126
- H 126, 104, 82, 22, 85, 147
- D 120, 135, 152, 116, 112, 63
- I 41, 151, 178, 103, 26, 148
- E 92, 104, 103, 129, 92, 43
- J 57, 110, 89, 162, 145, 113

DAY 25
- A 88, 125, 96, 130, 120, 149
- F 87, 123, 137, 135, 178, 126
- B 88, 98, 59, 106, 89, 99
- G 165, 114, 175, 93, 140, 128
- C 87, 87, 152, 91, 69, 132
- H 87, 66, 78, 137, 135, 86
- D 72, 149, 100, 43, 147, 134
- I 102, 183, 125, 87, 116, 115
- E 80, 67, 62, 110, 103, 94
- J 136, 95, 80, 150, 133, 114

DAY 26
- A 62, 107, 150, 126, 92, 54
- F 109, 95, 107, 124, 97, 142
- B 129, 84, 87, 38, 129, 141
- G 98, 62, 133, 148, 73, 103
- C 107, 112, 74, 109, 104, 59
- H 71, 136, 105, 114, 175, 80
- D 131, 114, 90, 138, 108, 62
- I 74, 120, 171, 85, 189, 125
- E 111, 69, 126, 46, 95, 73
- J 113, 79, 118, 97, 123, 160

DAY 27
- A 72, 83, 102, 163, 46, 151
- F 97, 95, 121, 79, 127, 141
- B 110, 88, 156, 122, 139, 98
- G 135, 97, 134, 37, 63, 55
- C 49, 82, 87, 152, 92, 181
- H 116, 156, 57, 99, 128, 64
- D 158, 31, 98, 134, 44, 131
- I 85, 131, 57, 87, 122, 57
- E 57, 110, 89, 162, 145, 113
- J 77, 179, 146, 82, 120, 167

DAY 28
- A 131, 89, 83, 60, 151, 25
- F 89, 107, 52, 147, 172, 111
- B 110, 122, 51, 54, 71, 145
- G 177, 89, 95, 98, 100, 108
- C 140, 42, 109, 75, 105, 104
- H 130, 157, 61, 142, 102, 60
- D 122, 91, 137, 140, 72, 144
- I 191, 97, 160, 110, 127, 137
- E 110, 46, 106, 130, 175, 56
- J 83, 106, 133, 85, 102, 153

DAY 29
- A 62, 60, 113, 142, 116, 120
- F 80, 159, 118, 131, 46, 118
- B 133, 121, 92, 61, 118, 126
- G 124, 101, 99, 70, 176, 91
- C 150, 88, 147, 47, 99, 129
- H 52, 113, 132, 82, 128, 136
- D 132, 67, 182, 57, 96, 67
- I 114, 152, 129, 196, 34, 103
- E 105, 152, 53, 117, 119, 89
- J 115, 164, 124, 162, 62, 185

DAY 30
- A 166, 120, 166, 142, 166, 69
- F 38, 104, 67, 100, 133, 96
- B 165, 89, 88, 90, 111, 129
- G 49, 131, 72, 156, 87, 25
- C 111, 114, 86, 130, 82, 187
- H 69, 69, 145, 29, 130, 134
- D 114, 55, 108, 85, 163, 124
- I 82, 76, 98, 95, 112, 111
- E 133, 116, 113, 52, 100, 58
- J 102, 46, 57, 63, 118, 118

DAY 31
- A 111, 80, 69, 114, 110, 115
- F 94, 130, 127, 97, 72, 143
- B 144, 120, 118, 103, 124, 121
- G 155, 133, 89, 62, 117, 69
- C 44, 76, 150, 95, 122, 119
- H 80, 98, 41, 90, 146, 27
- D 107, 49, 136, 140, 121, 107
- I 66, 109, 28, 183, 146, 170
- E 131, 74, 97, 79, 175, 129
- J 87, 126, 146, 164, 146, 145

DAY 32
- A 103, 90, 94, 101, 161, 106
- F 44, 126, 128, 93, 167, 115
- B 137, 61, 116, 81, 107, 153
- G 124, 59, 125, 39, 135, 105
- C 81, 116, 143, 131, 95, 111
- H 178, 112, 86, 132, 64, 138
- D 136, 122, 106, 108, 93, 84
- I 142, 86, 141, 64, 139, 85
- E 105, 80, 115, 82, 82, 111
- J 144, 109, 93, 142, 114, 161

DAY 33
- A 176, 144, 155, 144, 63, 189
- F 87, 127, 108, 76, 130, 80
- B 66, 165, 105, 158, 62, 117
- G 154, 101, 71, 75, 114, 158
- C 87, 66, 78, 137, 135, 86
- H 98, 37, 131, 70, 65, 71
- D 104, 107, 105, 107, 101, 111
- I 61, 121, 127, 90, 103, 94
- E 92, 73, 140, 89, 105, 130
- J 130, 80, 149, 145, 171, 153

DAY 34
- A 161, 64, 155, 101, 106, 36
- F 42, 59, 42, 84, 105, 61
- B 151, 183, 69, 113, 123, 106
- G 117, 98, 121, 93, 105, 69
- C 131, 96, 124, 111, 147, 149
- H 71, 127, 171, 170, 102, 84
- D 151, 105, 149, 150, 94, 137
- I 102, 106, 120, 124, 91, 94
- E 163, 66, 86, 54, 22, 65
- J 134, 23, 92, 68, 187, 95

DAY 35
- A 108, 112, 121, 47, 77, 114
- F 127, 154, 155, 140, 123, 72
- B 101, 62, 61, 96, 163, 65
- G 150, 116, 110, 171, 143, 121
- C 134, 70, 120, 55, 122, 129
- H 78, 76, 80, 89, 148, 26
- D 87, 81, 86, 63, 111, 111
- I 78, 194, 53, 167, 166, 110
- E 98, 83, 161, 129, 105, 109
- J 169, 44, 160, 178, 75, 52

DAY 36
- A 56, 129, 92, 162, 88, 72
- F 175, 155, 128, 34, 56, 115
- B 80, 133, 87, 123, 117, 103
- G 116, 129, 121, 177, 136, 167
- C 175, 123, 183, 113, 117, 114
- H 141, 157, 107, 145, 107, 124
- D 96, 155, 141, 107, 74, 58
- I 155, 129, 184, 67, 97, 111
- E 134, 171, 101, 119, 83, 128
- J 105, 120, 107, 123, 117, 141

DAY 37
- A 49, 76, 96, 120, 54, 90
- F 143, 157, 146, 116, 166, 97
- B 80, 41, 40, 154, 78, 71
- G 116, 129, 121, 177, 136, 167
- C 101, 60, 74, 131, 175, 92
- H 140, 112, 96, 87, 95, 63
- D 141, 120, 106, 93, 42, 52
- I 100, 50, 152, 93, 88, 166
- E 69, 106, 108, 108, 151, 153
- J 54, 144, 134, 101, 79, 39

DAY 38
- A 110, 70, 119, 86, 93, 130
- F 148, 77, 138, 104, 117, 82
- B 116, 147, 166, 93, 136, 120
- G 153, 87, 152, 77, 112, 102
- C 55, 85, 60, 160, 120, 126
- H 114, 115, 103, 34, 49, 115
- D 107, 121, 110, 122, 89, 167
- I 61, 175, 48, 58, 173, 170
- E 62, 112, 42, 45, 54, 76
- J 134, 96, 90, 109, 174, 76

DAY 39
- A 181, 75, 81, 154, 128, 43
- F 71, 157, 150, 114, 24, 91
- B 189, 117, 78, 119, 115, 185
- G 97, 180, 122, 152, 60, 176
- C 138, 119, 124, 156, 135, 89
- H 91, 128, 162, 131, 156, 50
- D 53, 71, 142, 83, 118, 105
- I 147, 126, 112, 130, 103, 43
- E 110, 149, 62, 35, 98, 186
- J 99, 75, 124, 45, 141, 162

DAY 40
- A 120, 166, 103, 92, 123, 121
- F 121, 107, 112, 61, 170, 110
- B 81, 61, 102, 104, 126, 114
- G 172, 41, 86, 161, 152, 134
- C 43, 136, 105, 95, 121, 143
- H 138, 132, 149, 108, 167, 167
- D 104, 71, 95, 113, 123, 147
- I 104, 137, 103, 79, 172, 58
- E 49, 85, 127, 142, 106, 110
- J 124, 86, 135, 64, 62, 161

DAY 41
- A 98, 105, 100, 61, 141, 95
- F 159, 100, 75, 109, 46, 85
- B 129, 124, 33, 147, 150, 31
- G 105, 190, 156, 102, 86, 166
- C 61, 150, 145, 78, 97, 112
- H 134, 126, 88, 150, 128, 117
- D 116, 142, 88, 95, 119, 87
- I 152, 163, 88, 90, 46, 79
- E 93, 127, 118, 155, 131, 128
- J 82, 95, 157, 101, 73, 82

DAY 42
- A 109, 93, 136, 98, 104, 118
- F 132, 142, 113, 157, 115, 102
- B 181, 127, 74, 111, 105, 147
- G 94, 107, 78, 42, 143, 71
- C 147, 62, 176, 195, 107, 50
- H 74, 163, 63, 98, 177, 62
- D 123, 139, 142, 105, 127, 161
- I 156, 164, 56, 135, 31, 82
- E 112, 131, 163, 95, 129, 43
- J 112, 94, 117, 58, 126, 76

DAY 43
- A 145, 136, 67, 139, 49, 83
- F 130, 88, 63, 51, 75, 95
- B 112, 32, 180, 51, 84, 50
- G 108, 114, 107, 97, 144, 69
- C 134, 196, 119, 78, 64, 62
- H 85, 123, 120, 82, 107, 115
- D 113, 106, 73, 153, 108, 54
- I 48, 108, 96, 117, 101, 77
- E 157, 69, 115, 78, 102, 130
- J 113, 81, 126, 136, 131, 85

DAY 44
- A 67, 164, 185, 61, 117, 166
- F 61, 57, 100, 101, 75, 95
- B 75, 114, 44, 82, 120, 165
- G 138, 138, 68, 125, 161, 160
- C 94, 113, 186, 63, 40, 119
- H 133, 109, 59, 67, 117, 131
- D 82, 49, 98, 40, 110, 35
- I 100, 150, 24, 106, 124, 87
- E 85, 129, 136, 109, 109, 115
- J 69, 129, 117, 100, 78, 140

DAY 45
- A 120, 101, 157, 103, 110, 49
- F 112, 170, 82, 160, 105, 98
- B 112, 113, 108, 139, 164, 105
- G 114, 102, 190, 130, 80, 146
- C 44, 80, 67, 148, 63, 58
- H 77, 118, 165, 94, 48, 118
- D 86, 106, 83, 132, 117, 162
- I 138, 137, 60, 126, 54, 119
- E 110, 122, 96, 177, 63, 132
- J 86, 70, 83, 93, 112, 105

DAY 46
- A 149, 145, 111, 111, 97, 154
- F 87, 46, 97, 92, 78, 105
- B 128, 138, 121, 64, 130, 86
- G 136, 82, 52, 137, 122, 103
- C 105, 85, 142, 120, 159, 50
- H 71, 150, 82, 72, 126, 91
- D 124, 102, 64, 142, 81, 129
- I 101, 85, 120, 63, 131, 84
- E 135, 116, 45, 138, 88, 153
- J 121, 105, 67, 78, 102, 177

DAY 47
- A 135, 146, 44, 180, 41, 148
- F 68, 56, 70, 167, 111, 168
- B 145, 104, 140, 152, 112, 114
- G 117, 170, 84, 81, 102, 123
- C 109, 124, 146, 91, 164, 137
- H 80, 117, 140, 100, 66, 135
- D 45, 118, 132, 83, 95, 128
- I 92, 48, 191, 80, 166, 39
- E 124, 118, 102, 82, 90, 95
- J 72, 99, 112, 80, 153, 151

DAY 48
- A 89, 151, 111, 145, 38, 95
- F 117, 41, 100, 106, 87, 81
- B 95, 114, 129, 121, 40, 82
- G 133, 43, 133, 169, 83, 149
- C 81, 81, 96, 129, 165, 157
- H 71, 178, 89, 122, 96, 88
- D 102, 104, 142, 46, 85, 129
- I 189, 122, 165, 140, 155, 158
- E 156, 109, 70, 132, 110, 54
- J 167, 74, 114, 61, 40, 139

DAY 49
- A 130, 102, 57, 117, 64, 153
- F 139, 81, 54, 30, 120, 53
- B 89, 150, 182, 116, 95, 94
- G 130, 28, 122, 110, 132, 95
- C 97, 108, 143, 120, 49, 95
- H 84, 152, 157, 121, 108, 190
- D 120, 95, 47, 92, 65, 51
- I 102, 95, 47, 92, 65, 51
- E 131, 89, 72, 149, 157, 39
- J 71, 123, 178, 124, 98, 146

DAY 50
- A 122, 94, 112, 93, 139, 99
- F 91, 151, 100, 139, 42, 196
- B 85, 150, 97, 121, 81, 98
- G 151, 131, 150, 74, 75, 70
- C 112, 120, 143, 131, 86, 79
- H 52, 110, 94, 122, 80, 164
- D 159, 142, 163, 100, 95, 142
- I 159, 142, 163, 100, 95, 142
- E 58, 130, 142, 73, 118, 102
- J 155, 72, 111, 145, 64, 69

DAY 51
- A 85, 78, 100, 74, 108, 134
- F 122, 85, 71, 30, 43, 62
- B 89, 150, 182, 116, 95, 94
- G 130, 28, 122, 110, 132, 95
- C 84, 152, 157, 121, 108, 190
- H 52, 110, 94, 122, 80, 164
- D 120, 95, 47, 92, 65, 51
- I 159, 142, 163, 100, 95, 142
- E 58, 130, 142, 73, 118, 102
- J 155, 72, 111, 145, 64, 69

DAY 52
- A 90, 87, 76, 106, 157, 51
- F 117, 103, 142, 102, 98, 119
- B 106, 54, 111, 90, 120, 84
- G 128, 74, 68, 105, 74, 59
- C 118, 61, 131, 91, 107, 89
- H 187, 137, 113, 79, 127, 48
- D 112, 71, 88, 95, 112, 100
- I 172, 132, 100, 147, 102, 98
- E 113, 162, 83, 101, 84, 99
- J 70, 144, 110, 108, 147, 122

DAY 53
- A 110, 99, 38, 179, 164, 122
- F 186, 96, 44, 108, 46, 142
- B 127, 81, 119, 74, 125, 114
- G 110, 57, 165, 70, 79, 64
- C 115, 135, 121, 43, 50, 134
- H 120, 160, 121, 43, 50, 134
- D 68, 71, 106, 111, 137, 118
- I 58, 128, 91, 140, 87, 76
- E 111, 104, 52, 105, 165, 193
- J 108, 114, 103, 77, 87, 164

DAY 54
- A 188, 125, 121, 125, 102, 42
- F 127, 183, 107, 106, 96, 152
- B 122, 184, 129, 43, 81, 69
- G 91, 68, 95, 113, 157, 94
- C 126, 78, 140, 145, 105, 132
- H 106, 58, 155, 145, 148, 97
- D 113, 57, 126, 98, 99, 155
- I 134, 108, 133, 192, 186, 147
- E 139, 75, 116, 121, 84, 175
- J 64, 117, 140, 130, 154, 72

DAY 55
- A 107, 126, 105, 55, 75, 90
- F 121, 118, 134, 64, 45, 144
- B 132, 130, 54, 113, 100, 124
- G 62, 98, 144, 87, 104, 124
- C 77, 125, 132, 150, 56, 65
- H 108, 97, 61, 52, 111, 155
- D 95, 63, 53, 109, 114, 179
- I 114, 70, 146, 170, 143, 89
- E 37, 114, 170, 129, 73, 85
- J 147, 169, 102, 164, 128, 74

Answer Key Mapping:
Day 1 Problems & Solutions

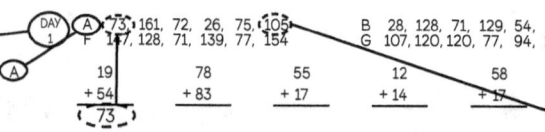

DAY 1
- A 73, 161, 72, 26, 75, 105
- F 147, 128, 71, 139, 77, 154
- B 28, 128, 71, 129, 54, 95
- G 107, 120, 120, 77, 94, 102

19	78	55	12	58	11
+ 54	+ 83	+ 17	+ 14	+ 17	+ 94
73					105

Two rows (A–J) with six columns of solutions for each day.

Subtraction Answer Key Sheet

Block A / F

Day	Row						
1	A	31	19	58	34	29	40
1	F	2	51	9	12	19	66
2	A	41	0	55	31	13	47
2	F	49	22	38	55	2	5
3	A	50	40	78	27	22	52
3	F	33	56	45	0	5	52
4	A	20	16	70	20	70	0
4	F	5	30	44	50	1	2
5	A	10	32	22	46	65	40
5	F	29	42	0	59	18	4
6	A	3	6	4	12	22	17
6	F	36	7	62	9	8	64
7	A	70	12	73	10	7	9
7	F	16	73	17	67	21	7
8	A	7	56	80	43	36	57
8	F	20	4	35	77	33	1
9	A	6	16	20	34	26	57
9	F	2	38	4	3	26	17
10	A	43	10	62	7	4	38
10	F	63	36	43	10	28	13
11	A	1	8	61	41	69	42
11	F	65	46	45	14	66	8
12	A	2	77	25	25	9	22
12	F	6	43	48	14	50	3
13	A	2	11	44	11	26	55
13	F	68	27	5	20	20	2
14	A	15	5	2	4	12	22
14	F	3	37	20	37	53	52
15	A	41	53	24	11	33	42
15	F	15	13	25	12	39	13
16	A	36	10	9	46	56	63
16	F	69	45	3	25	45	10
17	A	15	30	29	29	23	16
17	F	26	21	54	23	32	6
18	A	2	53	68	29	6	71
18	F	3	57	5	18	3	24
19	A	15	25	17	14	28	27
19	F	39	54	41	25	7	38
20	A	32	20	1	15	48	75
20	F	58	12	11	53	53	10
21	A	39	18	44	82	13	8
21	F	46	2	26	1	40	81
22	A	27	15	39	40	5	34
22	F	0	21	32	38	37	1
23	A	25	10	1	23	31	19
23	F	10	6	12	67	10	2
24	A	21	19	23	1	33	63
24	F	32	43	44	14	42	30
25	A	9	15	22	44	2	12
25	F	32	32	25	44	5	16
26	A	32	58	26	6	16	35
26	F	24	45	1	8	22	40
27	A	16	40	65	20	5	48
27	F	22	11	12	6	10	5
28	A	9	60	33	47	14	1
28	F	17	61	2	48	21	6
29	A	43	24	20	77	55	28
29	F	53	14	65	10	5	14
30	A	72	35	35	28	17	23
30	F	37	9	75	8	3	38
31	A	74	26	78	38	34	19
31	F	17	52	14	27	65	8
32	A	61	3	25	81	61	77
32	F	49	7	16	4	36	11
33	A	42	68	42	60	29	71
33	F	11	6	25	79	1	64
34	A	18	18	13	38	28	28
34	F	6	30	5	14	33	1
35	A	19	41	29	8	0	3
35	F	23	18	11	23	70	32
36	A	6	35	14	57	68	25
36	F	17	12	11	5	25	41
37	A	9	2	77	40	25	12
37	F	18	35	22	41	3	16
38	A	8	59	27	19	2	59
38	F	6	12	67	29	29	1
39	A	9	5	54	1	38	45
39	F	5	61	15	6	80	0
40	A	2	61	54	34	32	31
40	F	47	29	35	14	10	68
41	A	13	21	21	3	3	43
41	F	1	14	37	50	30	42
42	A	74	49	2	32	7	36
42	F	6	52	25	69	4	7
43	A	74	40	34	46	28	28
43	F	47	42	40	39	55	19
44	A	34	2	33	77	38	33
44	F	26	4	15	3	3	3
45	A	55	59	46	19	33	11
45	F	54	35	57	63	56	54
46	A	56	56	65	72	5	24
46	F	46	3	36	53	32	64
47	A	16	26	26	39	39	39
47	F	59	17	45	1	7	73
48	A	74	64	36	18	21	52
48	F	15	41	21	16	18	35
49	A	45	8	51	31	6	55
49	F	49	71	26	71	15	10
50	A	44	27	12	43	27	17
50	F	57	31	32	20	72	14
51	A	25	5	29	42	6	2
51	F	32	22	16	20	16	41
52	A	47	36	21	17	4	19
52	F	62	17	9	29	0	77
53	A	15	16	19	26	28	5
53	F	23	12	25	18	23	6
54	A	34	64	67	12	10	26
54	F	2	16	30	24	9	78
55	A	3	16	30	24	9	78
55	F	16	5	24	5	33	34

Block B / G

Day	Row						
1	B	18	14	47	44	57	8
1	G	23	65	6	9	27	20
2	B	6	19	5	31	4	65
2	G	3	18	23	49	45	3
3	B	32	12	20	20	25	11
3	G	45	61	22	62	9	76
4	B	18	20	3	63	31	8
4	G	70	26	14	67	55	8
5	B	3	27	38	76	43	54
5	G	22	0	22	41	1	39
6	B	9	3	8	36	7	8
6	G	3	5	3	32	27	21
7	B	32	34	34	18	42	51
7	G	12	31	30	57	44	65
8	B	60	33	1	9	28	2
8	G	32	32	10	17	19	21
9	B	12	41	34	51	10	30
9	G	32	12	23	10	56	11
10	B	16	51	12	3	9	30
10	G	82	42	54	42	26	45
11	B	17	24	8	26	41	8
11	G	35	3	18	7	35	14
12	B	22	59	43	70	6	11
12	G	31	23	9	3	19	3
13	B	68	20	35	23	5	33
13	G	2	25	61	22	19	18
14	B	58	32	50	10	5	21
14	G	5	4	7	18	62	33
15	B	64	16	5	60	31	1
15	G	37	24	29	56	13	36
16	B	29	58	5	9	13	36
16	G	42	18	13	39	7	1
17	B	3	75	68	80	30	47
17	G	40	75	82	34	13	30
18	B	21	25	14	3	37	19
18	G	30	51	4	24	19	60
19	B	12	50	27	71	61	28
19	G	60	65	80	10	21	9
20	B	3	56	7	20	25	7
20	G	75	22	10	18	30	64
21	B	69	26	23	32	24	7
21	G	49	18	14	31	40	4
22	B	1	55	48	45	48	28
22	G	20	37	37	5	32	2
23	B	18	55	40	28	67	1
23	G	81	53	22	4	55	9
24	B	25	55	0	39	25	24
24	G	56	20	2	35	17	42
25	B	11	51	7	2	17	28
25	G	80	56	52	47	4	4
26	B	52	8	17	5	19	37
26	G	15	0	14	47	64	83
27	B	42	18	22	15	7	43
27	G	35	4	35	72	77	44
28	B	57	19	35	23	10	20
28	G	3	10	5	27	62	17
29	B	9	17	51	8	28	67
29	G	8	15	13	78	4	12
30	B	24	41	16	20	21	25
30	G	15	47	35	59	23	4
31	B	47	7	22	34	15	50
31	G	7	1	9	53	84	50
32	B	29	48	23	41	57	20
32	G	18	22	6	14	18	52
33	B	8	8	45	64	51	11
33	G	34	1	6	2	17	23
34	B	51	21	13	19	56	32
34	G	49	25	46	47	17	30
35	B	41	28	42	43	6	12
35	G	62	42	18	32	16	18
36	B	9	25	11	6	20	53
36	G	25	5	63	28	79	30
37	B	21	14	21	45	12	57
37	G	3	68	1	23	45	41
38	B	40	30	28	41	25	16
38	G	12	2	42	18	40	72
39	B	14	11	14	52	10	21
39	G	3	1	14	4	18	40
40	B	23	29	16	14	6	14
40	G	3	28	25	32	77	1
41	B	65	8	21	25	21	27
41	G	28	2	43	29	45	31
42	B	28	59	36	56	16	32
42	G	13	57	44	1	18	13
43	B	58	27	48	11	74	45
43	G	2	50	20	39	51	3
44	B	36	55	62	26	13	3
44	G	11	41	28	19	8	9
45	B	42	10	47	54	15	41
45	G	44	26	40	5	27	68
46	B	16	38	41	58	48	7
46	G	12	75	17	17	52	1
47	B	14	6	23	13	10	45
47	G	62	25	35	22	1	32
48	B	4	58	55	2	16	1
48	G	17	4	68	33	11	60
49	B	21	5	43	55	65	65
49	G	18	23	16	14	13	0
50	B	32	7	6	69	56	26
50	G	54	29	9	6	22	55
51	B	13	36	6	22	55	33
51	G	51	34	35	8	3	10
52	B	6	69	67	21	55	11
52	G	26	27	59	6	36	37
53	B	5	16	16	36	44	54
53	G	44	62	3	8	5	5
54	B	17	8	19	64	6	4
54	G	30	11	38	4	67	55

Block C / H

Day	Row						
1	C	46	11	75	18	32	20
1	H	4	57	2	25	2	5
2	C	26	50	32	87	1	23
2	H	13	4	31	50	25	0
3	C	50	15	38	48	58	13
3	H	19	34	47	48	69	40
4	C	28	54	73	9	81	13
4	H	63	1	65	57	36	48
5	C	35	4	59	4	18	34
5	H	54	46	7	5	2	34
6	C	58	2	47	61	83	10
6	H	22	26	1	12	38	33
7	C	1	61	56	6	7	30
7	H	39	41	29	16	65	39
8	C	19	65	31	30	25	17
8	H	7	49	63	41	39	4
9	C	52	12	59	72	31	27
9	H	2	20	42	37	7	22
10	C	40	26	43	21	34	1
10	H	19	20	11	8	40	43
11	C	27	18	61	24	64	73
11	H	55	45	26	23	3	69
12	C	36	2	29	68	63	36
12	H	55	33	57	56	28	3
13	C	73	14	0	0	23	33
13	H	2	41	8	11	6	46
14	C	2	19	24	14	9	29
14	H	31	14	14	41	74	32
15	C	2	2	18	11	18	21
15	H	31	1	4	15	48	8
16	C	16	21	23	76	42	4
16	H	43	1	5	1	5	82
17	C	59	5	20	38	14	46
17	H	32	5	33	12	17	13
18	C	26	32	34	49	50	63
18	H	26	10	58	13	45	63
19	C	24	46	7	27	48	76
19	H	59	52	26	9	30	4
20	C	14	39	9	19	64	60
20	H	12	10	21	26	67	78
21	C	31	13	23	4	58	45
21	H	18	79	21	7	28	37
22	C	24	35	46	49	60	8
22	H	54	40	73	15	53	5
23	C	10	43	38	53	24	16
23	H	52	25	37	53	28	22
24	C	21	46	6	24	70	40
24	H	15	35	16	33	28	37
25	C	9	9	30	37	3	30
25	H	20	51	18	76	19	70
26	C	13	7	56	16	39	32
26	H	19	41	47	54	2	17
27	C	53	25	8	41	79	20
27	H	5	76	28	4	5	34
28	C	60	41	37	10	7	23
28	H	9	47	69	25	1	37
29	C	14	32	42	26	16	21
29	H	53	57	6	29	13	37
30	C	2	34	52	56	33	3
30	H	35	12	10	9	24	25
31	C	1	37	47	17	7	20
31	H	19	12	50	27	36	71
32	C	12	21	3	51	10	40
32	H	27	16	39	41	61	41
33	C	71	43	41	43	59	25
33	H	28	66	3	20	17	18
34	C	28	15	4	26	23	23
34	H	25	9	34	42	12	15
35	C	15	10	15	43	12	41
35	H	48	36	40	73	21	39
36	C	31	49	23	39	72	62
36	H	3	53	27	60	49	61
37	C	52	4	58	32	4	42
37	H	39	13	7	22	33	24
38	C	59	29	58	8	41	18
38	H	28	69	11	36	61	46
39	C	3	5	15	44	6	5
39	H	11	42	6	31	32	30
40	C	66	51	30	50	33	19
40	H	69	4	29	7	40	8
41	C	5	8	50	14	8	58
41	H	6	47	30	17	50	6
42	C	2	0	35	10	21	12
42	H	32	27	48	18	14	17
43	C	9	43	9	34	17	56
43	H	33	37	9	3	26	1
44	C	55	41	79	32	23	19
44	H	35	81	51	44	57	17
45	C	13	26	18	61	39	43
45	H	35	61	71	9	30	66
46	C	0	3	16	15	34	6
46	H	13	4	27	22	21	71
47	C	35	45	15	19	43	6
47	H	14	14	42	74	59	23
48	C	23	28	41	2	29	6
48	H	79	18	78	16	41	17
49	C	37	30	23	36	20	6
49	H	17	19	0	66	19	16
50	C	7	8	12	35	7	31
50	H	34	83	12	13	2	1
51	C	50	6	5	19	78	38
51	H	39	51	53	7	64	29
52	C	74	35	8	58	3	10
52	H	1	12	28	16	13	2
53	C	61	3	32	37	8	15
53	H	38	59	22	47	22	23
54	C	36	35	37	20	22	3
54	H	67	3	12	9	53	21
55	H	3	17	8	15	52	9

Block D / I

Day	Row						
1	D	42	59	19	22	24	51
1	I	22	25	12	35	16	11
2	D	2	38	8	42	12	16
2	I	1	28	26	2	64	31
3	D	21	14	33	0	2	27
3	I	60	55	29	70	61	43
4	D	17	11	2	28	9	45
4	I	68	39	45	48	36	83
5	D	51	38	22	47	23	7
5	I	33	28	28	17	39	28
6	D	13	7	22	70	7	9
6	I	34	39	3	22	58	14
7	D	39	18	26	39	42	12
7	I	6	43	37	1	36	1
8	D	28	40	12	15	46	63
8	I	58	24	13	22	1	14
9	D	20	73	1	55	47	18
9	I	76	4	41	11	45	12
10	D	2	1	2	12	2	2
10	I	2	27	37	65	17	26
11	D	30	5	44	56	10	31
11	I	10	1	19	3	7	17
12	D	14	37	72	37	34	72
12	I	40	52	26	22	23	34
13	D	16	59	68	39	81	6
13	I	39	1	8	27	37	59
14	D	11	71	14	4	8	49
14	I	9	6	4	22	74	10
15	D	60	64	43	9	3	25
15	I	31	17	43	32	60	36
16	D	76	21	62	2	5	39
16	I	3	46	4	13	43	72
17	D	79	21	77	78	55	26
17	I	49	83	7	3	48	40
18	D	24	17	40	34	56	3
18	I	35	6	6	10	47	25
19	D	47	24	3	44	36	7
19	I	7	45	16	36	22	42
20	D	16	62	30	29	1	5
20	I	42	21	52	37	11	17
21	D	8	11	47	11	3	19
21	I	56	48	8	20	19	32
22	D	77	55	76	10	46	42
22	I	6	16	16	6	6	24
23	D	67	18	38	21	39	1
23	I	5	69	57	12	3	5
24	D	65	4	74	46	4	39
24	I	55	24	46	20	44	22
25	D	46	15	6	45	26	39
25	I	22	11	65	11	30	47
26	D	52	14	2	11	7	21
26	I	44	45	19	66	38	15
27	D	22	14	9	8	38	25
27	I	26	0	66	34	29	12
28	D	20	21	62	39	33	62
28	I	33	40	1	29	44	84
29	D	45	3	16	42	14	73
29	I	21	76	7	84	31	28
30	D	40	15	46	33	47	24
30	I	26	55	34	11	15	81
31	D	6	39	52	14	9	45
31	I	18	17	43	32	69	34
32	D	29	3	12	37	46	63
32	I	64	55	40	1	18	12
33	D	36	12	46	18	10	2
33	I	3	25	54	15	57	7
34	D	5	16	10	32	50	45
34	I	40	66	19	26	12	36
35	D	24	20	17	27	18	18
35	I	29	28	84	23	81	40
36	D	67	49	19	11	1	34
36	I	34	22	14	65	74	47
37	D	20	68	9	6	20	46
37	I	25	24	0	37	72	52
38	D	13	14	7	2	35	36
38	I	62	31	64	27	13	56
39	D	8	10	5	30	2	31
39	I	8	37	11	43	71	6
40	D	65	29	31	23	31	50
40	I	46	29	1	42	32	38
41	D	28	15	11	47	6	45
41	I	66	10	28	10	32	14
42	D	18	31	44	27	35	1
42	I	64	1	1	30	13	6
43	D	1	9	34	37	56	51
43	I	9	2	30	70	52	0
44	D	37	6	41	1	30	10
44	I	20	8	72	14	2	10
45	D	79	18	78	16	41	1
45	I	25	52	20	36	20	19
46	D	14	50	45	32	25	9
46	I	55	15	51	17	51	1
47	D	1	9	34	37	56	51
47	I	27	26	1	57	1	8
48	D	37	6	41	1	30	10
48	I	1	69	43	8	45	1
49	D	34	71	29	26	9	50
49	I	81	13	31	20	71	
50	D	7	34	70	78	39	63
50	I	7	39	42	15	6	30
51	D	38	15	67	9	18	9
51	I	65	14	44	6	48	46
52	D	26	30	3	4	8	36
52	I	27	58	35	30	0	10
53	D	6	5	43	21	41	41
53	I	42	65	50	43	34	10
54	D	2	29	2	20	4	35
54	I	26	53	13	6	34	13

Block E / J

Day	Row						
1	E	1	35	45	34	12	21
1	J	49	32	23	22	48	26
2	E	63	23	33	59	45	33
2	J	37	57	47	53	23	8
3	E	39	81	5	39	13	16
3	J	4	14	14	9	69	5
4	E	31	5	20	46	8	87
4	J	63	5	6	14	3	7
5	E	18	4	16	45	52	11
5	J	42	5	65	24	54	44
6	E	1	9	11	41	9	42
6	J	28	53	19	77	27	27
7	E	56	81	4	28	7	18
7	J	24	6	62	55	0	16
8	E	72	10	24	15	44	8
8	J	3	1	23	25	64	31
9	E	11	4	7	1	4	11
9	J	80	50	5	17	21	44
10	E	32	63	6	47	23	14
10	J	1	28	4	32	18	39
11	E	22	73	36	24	30	16
11	J	70	19	46	29	56	18
12	E	45	22	73	21	13	22
12	J	10	37	6	9	15	59
13	E	17	5	61	41	17	52
13	J	56	26	5	16	45	39
14	E	30	27	18	70	7	31
14	J	40	19	17	15	32	0
15	E	50	23	56	6	8	9
15	J	22	12	43	31	1	79
16	E	11	36	21	54	2	11
16	J	62	18	33	22	69	40
17	E	42	62	4	29	37	16
17	J	3	3	67	34	49	9
18	E	48	53	69	65	12	44
18	J	21	44	48	49	24	62
19	E	44	29	49	17	40	57
19	J	45	16	50	50	18	39
20	E	11	30	6	2	3	26
20	J	9	9	0	0	56	8
21	E	17	9	4	36	57	53
21	J	17	45	43	7	8	14
22	E	21	3	19	1	3	3
22	J	74	30	33	47	56	21
23	E	18	11	16	16	4	7
23	J	30	18	16	27	72	6
24	E	23	2	12	9	23	20
24	J	30	26	3	40	35	61
25	E	70	14	12	21	67	49
25	J	38	72	11	15	39	11
26	E	1	16	11	62	12	75
26	J	7	6	62	30	33	38
27	E	50	29	6	9	69	34
27	J	46	31	1	2	7	28
28	E	2	1	24	61	57	39
28	J	9	30	20	21	68	1
29	E	21	7	23	37	39	40
29	J	13	72	26	59	11	18
30	E	23	6	4	65	60	27
30	J	57	85	23	34	15	7
31	E	48	61	50	3	18	51
31	J	21	64	6	8	35	73
32	E	50	18	54	58	17	17
32	J	22	22	11	5	56	6
33	E	26	53	57	17	25	9
33	J	18	66	19	21	54	20
34	E	50	47	56	43	10	7
34	J	35	18	39	4	9	40
35	E	42	23	64	28	3	65
35	J	26	3	33	4	0	1
36	E	48	18	15	24	1	3
36	J	16	35	4	21	13	8
37	E	53	65	0	46	5	16
37	J	64	56	3	29	43	50
38	E	16	33	8	0	72	1
38	J	17	1	20	12	30	38
39	E	9	0	28	20	31	23
39	J	5	8	32	60	38	24
40	E	63	15	22	40	14	37
40	J	15	17	1	11	2	35
41	E	60	44	6	66	49	49
41	J	7	57	14	35	17	21
42	E	12	22	37	39	25	56
42	J	30	28	37	73	10	5
43	E	47	35	24	31	50	23
43	J	37	55	28	39	44	55
44	E	76	39	2	7	65	1
44	J	12	42	32	8	75	9
45	E	25	13	17	74	56	2
45	J	7	6	10	37	80	49
46	E	29	33	62	60	25	11
46	J	66	39	41	19	15	38
47	E	3	1	10	44	19	11
47	J	48	81	37	10	51	16
48	E	31	36	71	16	3	27
48	J	4	7	40	24	30	1
49	E	34	52	8	22	14	44
49	J	11	51	51	42	29	8
50	E	14	79	7	53	20	1
50	J	34	42	13	10	31	52
51	E	24	13	9	22	40	6
51	J	30	48	35	6	67	51

Answer Key Mapping

Day 1 Problems & Solutions

| | DAY 1 A | 31 | 19 | 58 | 34 | 29 | **40** | B | 18 | 14 | 47 | 44 | 57 | 8 |
| | F | 2 | 51 | 9 | 12 | 19 | 66 | G | 23 | 65 | 6 | 9 | 27 | 20 |

```
   83      72      82      49      73      68
 - 52    - 53    - 24    - 15    - 44    - 28
 ----    ----    ----    ----    ----    ----
 (31)                                    (40)
```

Two rows (A–J) with six columns of solutions for each day.

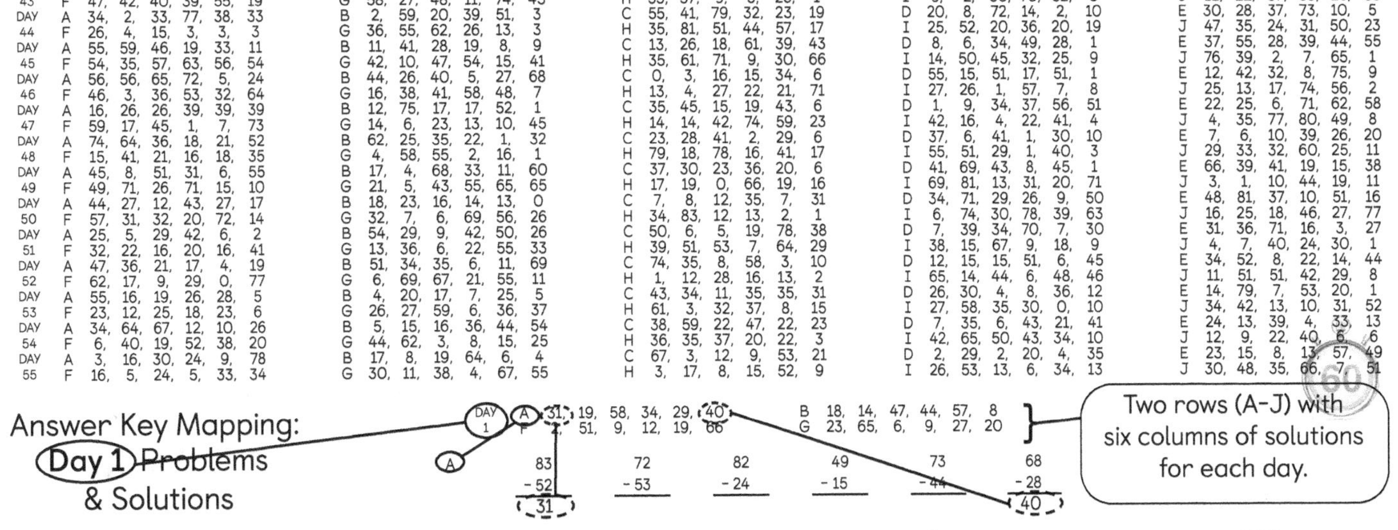

Check out our other books!

abcZbook Press

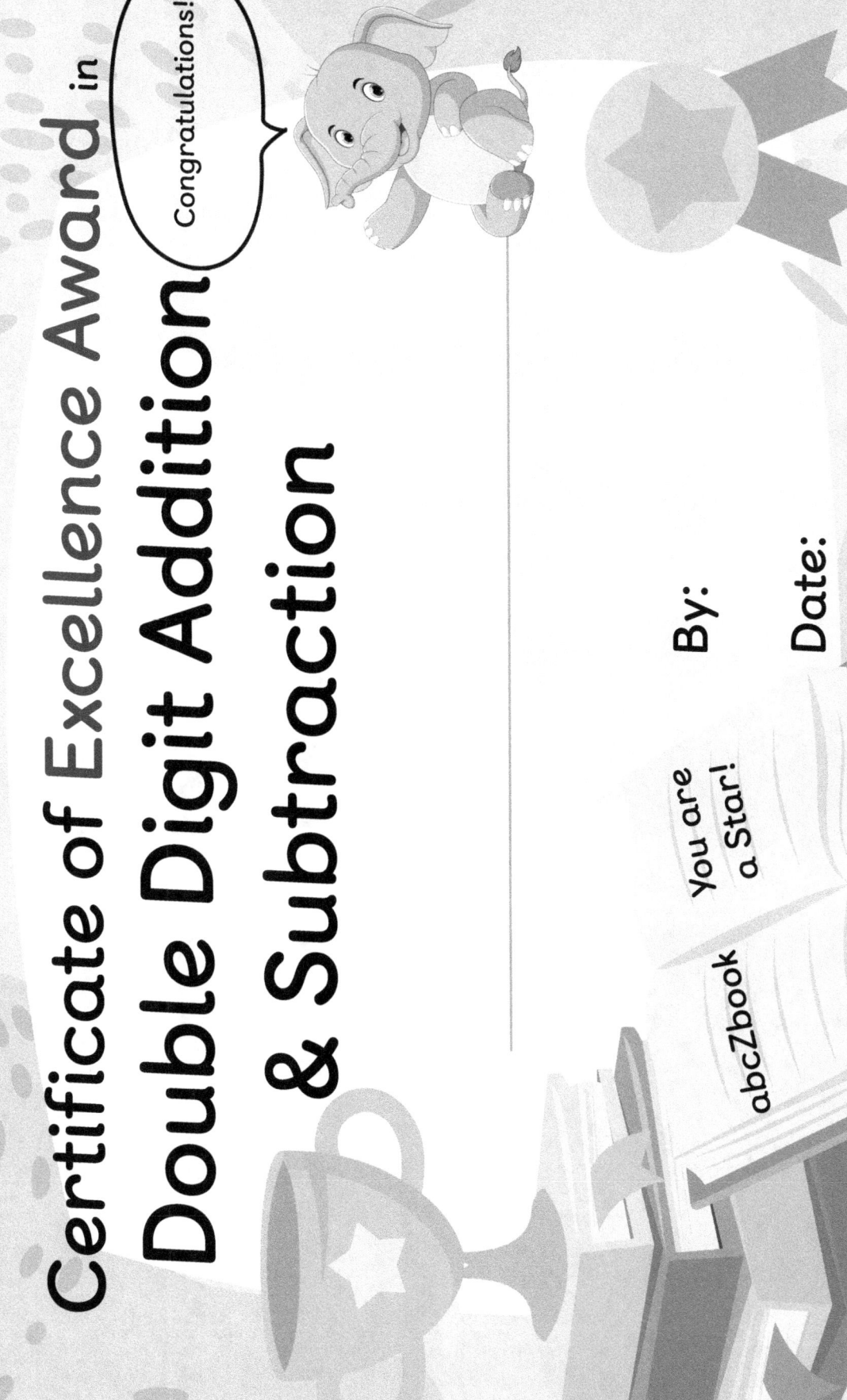

Certificate of Excellence Award in

Double Digit Addition & Subtraction

Congratulations!

You are a Star!

abcZbook

By:

Date: